Beijing Local Standard

Design Standard of Green Building

DB 11/938 — 2012

Chief Development Organizations: Beijing Municipal Office of Investigation & Design and
Surveying & Mapping Management
China Academy of Building Research
Tsinghua University
Approval Department: Beijing Municipal Commission of Urban Planning
Beijing Municipal Administration of Quality and Technology Supervision
Implementation Date: Jul. 1st, 2013

China Architecture & Building Press

Beijing 2014

图书在版编目(CIP)数据

绿色建筑设计标准 DB 11/938—2012/北京市勘察设计与测绘管理办公室组织编译. —北京：中国建筑工业出版社，2015.2
（工程建设标准英文版）
ISBN 978-7-112-17400-3

Ⅰ.①绿… Ⅱ.①北… Ⅲ.①生态建筑-建筑设计-标准-北京市-指南-英文 Ⅳ.①TU2-65

中国版本图书馆 CIP 数据核字(2014)第 264606 号

Chinese edition first published in the People's Republic of China in 2013
English edition first published in the People's Republic of China in 2015
by China Architecture & Building Press
No. 9 Sanlihe Road
Beijing, 100037
www. cabp. com. cn

Printed in China by BeiJing YanLinJiZhao printing CO. , LTD

© 2013 by Beijing Municipal Office of Investigation & Design and
Surveying & Mapping Management

All rights reserved. No part of this publication may be reproduced or transmitted in any form or
by any means, graphic, electronic, or mechanical, including photocopying, recording,
or any information storage and retrieval systems, without written permission of the publisher.

This book is sold subject to the condition that it shall not, by way of trade or otherwise, be lent,
re-sold, hired out or otherwise circulated without the publisher's prior consent in any form of
blinding or cover other than that in which this is published and without a similar condition
including this condition being imposed on the subsequent purchaser.

ISBN 978-7-112-17400-3(26165)

Announcement of Beijing Local Standard

NO. 11 in 2012 standard number (The total No. 124th)

The following one Beijing Local Standard Jointly Issued by Beijing Municipal Commission of Urban Planning and Beijing Municipal Administration of Quality and Technology Supervision, which has approved by Beijing Municipal Administration of Quality and Technology Supervision, and are made public(See appendix).

Appendix: Approval of the publication of Beijing Local Standard.

Beijing Municipal Commission of Urban Planning
Beijing Municipal Administration of Quality
and Technology Supervision
Dec. 12th, 2012

Appendix:

Approval of the publication of Beijing Local Standard

No.	No. of Local Standard	Name of Local Standard	Approval Data	Implementation Date
1	DB11/938—2012	Design Standard Of Green Buildings	Dec. 12th, 2012	Jul. 1st, 2013

Note: The above Local Standard in text can inquired by sign in the website of Beijing Municipal Administration of Quality and Technology Supervision (www. bjtsb. gov. cn) or website of the capital standard (www. china-std. com).

Foreword

This standard is formulated according to the standard development plan of Beijing Municipal Administration of Quality and Technology Supervision and Beijing Municipal Commission of Urban Planning and jointly by China Academy of Building Research, Tsinghua University and other units which have made a wide investigation and study, carefully summarized the practical experience in green buildings in recent years, referenced domestic and foreign associated standards and application study results, and combined the urban-rural construction and development demands of Beijing.

This standard includes totally 14 chapters, and their main technical contents are: 1. General Provisions; 2. Terms; 3. Basic Requirements; 4. Index System; 5. Scheme and Design Documents of Green Building; 6. Planning and Design; 7. Architectural Design; 8. Structure Design; 9. Design of Water Supply and Sewerage; 10. Heating, Ventilation and Air-conditioning Design; 11. Building Electrical Design; 12. Landscape Design; 13. Interior Decoration Design; and 14. Special Design Control.

The provisions printed in bold type are compulsory ones and must be enforced strictly.

This standard is put under the centralized management by Beijing Municipal Commission of Urban Planning. China Academy of Building Research and Tsinghua University are responsible for interpretation of its specific technical contents, and the Standardization Office of Urban-Rural Planning of Beijing is in charge of its daily management.

In order to adapt this standard to the needs of Beijing's green building design, all relevant organizations are kindly requested to feed back the comments and suggestions duly to China Academy of Building Research (address: No. 30, East Road, North 3rd Ring, Beijing; zip code: 100013; telephone: 64517259; Email: bjlsjzbz@cabr-design.com).

Telephone of Standardization Office of Urban-Rural Planning of Beijing: 68017520; Email: bjbb3000@163.com.

Chief Development Organizations:
      Beijing Municipal Office of Investigation & Design and Surveying & Mapping Management
      China Academy of Building Research
      Tsinghua University

Participating Development Organizations:
      Architectural Design and Research Institute of Tsinghua University Co., Ltd.
      Beijing Municipal Institute of City Planning & Design
      Beijing Institute of Architectural Design Co., Ltd.
      Zhongtian Weiye (Beijing) Architectural Design ASSOCIATES Co., Ltd.
      China IPPR International Engineering Corporation
      GWS - Green World Solutions
      College of Architecture and Urban Planning of Beijing University Of Technology

China Institute of Building Standard Design Research
Beijing QiDi Daring Energy Technology Co., Ltd.
Beijing Association for Engineering Construction Standardization
Shanghai Zero Carbon Information Technology Center

Participating Organizations:

Beijing Guoao Shidai New Energy Technology Development Co., Ltd.
Beijing Vanke Co., Ltd.

Chief Drafting Staff:

Ye Dahua   Zeng Jie   Ye Jia   Luo Wei
Zhu Yingxin   Zeng Yu   Ju Pengyan   Lin Borong
Huang Xianming   Jiao Jian   Xue Shiyong   Zhang Tongyi
Zhao Yange   Li Jianlin   Sheng Xiaokang   Wu Yan
Xu He   Huang Ning   Chen Zhe   Li Benqiang
Xia Wei   Liu Jiagen   Wan Shuie   Yang Hong
Chu Xin   Liu Huimin   Shen Hongming   Yu Qi
Hu Qian   Xiang Weizhong   Chen Shuo   Liu Yonghui   Robert Vale

Chief Reviewers:

Wu Desheng   Li Dexiang   Cheng Dazhang   Che Wu
Zhao Li   Zhang Bo   Wang Changxing

Contents

1 General Provisions .... (1)
2 Terms .... (2)
3 Basic Requirements .... (3)
4 Index System .... (4)
    4.1 General Requirements .... (4)
    4.2 Low Carbon Ecological Design Index System at Detailed Planning Stage .... (4)
    4.3 Green Design Index System at Building Design Stage .... (10)
5 Scheme and Design Documents of Green Building .... (15)
    5.1 Scheme of Green Building .... (15)
    5.2 Organization of Green Building Design .... (16)
    5.3 Design Documents of Green Building Design .... (16)
6 Planning and Design .... (18)
    6.1 General Requirements .... (18)
    6.2 Land Use Planning .... (18)
    6.3 Transportation Planning .... (20)
    6.4 Resource Utilization .... (21)
    6.5 Ecological Environment .... (22)
7 Architectural Design .... (25)
    7.1 General Requirements .... (25)
    7.2 Building Space Layout .... (25)
    7.3 Building Envelope .... (26)
    7.4 Building Material .... (27)
    7.5 Building Acoustical Environment .... (28)
    7.6 Daylighting .... (29)
    7.7 Natural Ventilation Environment .... (30)
    7.8 Indoor Air Quality .... (31)
    7.9 Other Requirements .... (31)
8 Structure Design .... (32)
    8.1 General Requirements .... (32)
    8.2 Design of Main Structure .... (32)
    8.3 Design of Building Foundation .... (33)
    8.4 Design of Renovation and Extension Structure .... (34)
9 Design of Water Supply and Sewerage .... (35)
    9.1 General Requirements .... (35)
    9.2 Water Supply System .... (35)
    9.3 Water Saving Equipment and Sanitary Ware .... (37)
    9.4 Utilization of Non-traditional Water Source .... (37)

10	Heating, Ventilation and Air-conditioning Design	(39)
10.1	General Requirements	(39)
10.2	Energy Transportation and Distribution	(40)
10.3	Heat and Cold Source	(42)
10.4	Control and Detection	(43)
11	Building Electrical Design	(45)
11.1	General Requirements	(45)
11.2	Power Supply and Distribution System	(45)
11.3	Lighting	(46)
11.4	Electrical Equipment	(47)
11.5	Metering and Intelligentization	(47)
12	Landscape Design	(50)
12.1	General Requirements	(50)
12.2	Greening	(50)
12.3	Waterscape	(51)
12.4	Sites	(52)
12.5	Lighting	(53)
13	Interior Decoration Design	(54)
13.1	General Requirements	(54)
13.2	Design Requirements	(54)
13.3	Selection of Decoration Materials	(55)
14	Special Design Control	(56)
14.1	General Requirements	(56)
14.2	Building Curtain Wall	(56)
14.3	Reclaimed water Treatment and Rainwater Recycling Systems	(56)
14.4	Solar Thermal and Photovoltaic Systems	(57)
14.5	Heat Pump System	(58)
14.6	Ice Storage System	(59)
14.7	Building Intelligent Systems	(59)
Appendix A	Integrated Review Sheet	(60)
Appendix B	Beijing Design Data Collection	(70)
Appendix C	Boundary Conditions for Simulation	(83)
Explanation of Wording in This Standard		(92)
List of Quoted Standards		(93)

1 General Provisions

1.0.1 This standard is formulated with a view to implementing the development strategy of "Culture-enriched Beijing, Technology-empowered Beijing and Environment-friendly Beijing" proposed by the People's Government of Beijing Municipality, and lead the development of low-carbon ecological planning and green buildings.

1.0.2 This standard is applicable to the green design and management of construction, renovation and extension buildings, and also to the low-carbon ecological planning at the detailed planning stage.

1.0.3 Green design of civil buildings shall take full consideration of the dialectical relationship of the building functions with energy, land, water and material saving and environmental protection in the building life cycle, and reflect the unity of economic, social and environmental benefit; reduce the impact of construction behaviors on the natural environment, comply with the healthy, simple and efficient design idea, and realize the harmonious coexistence of man, buildings and nature.

1.0.4 The green design of civil buildings and the low-carbon ecological planning shall comply with the requirements of the current national and Beijing's associated standards in addition to those of this standard.

2 Terms

2.0.1 Green design of civil buildings

Reflecting the idea of sustainable development in design, realizing resource saving and environmental protection in the building life cycle on the basics of meeting building functions, and providing the healthy, applicable and efficient use space for people.

2.0.2 Low-carbon ecological planning

Fusing the low-carbon target with the ecological idea, and realizing the harmonious coexistence of man and natural environment with the sustainable development by integrating the low-carbon ecological strategies of land, industry, traffic, energy, resource, environment and society in the regional, urban or block planning.

2.0.3 Building life cycle

All process of buildings from construction, use to demolition, including the acquisition of raw materials, the processing and manufacturing of building materials and components, the construction and installation on site, the operation and maintenance of buildings, and the final demolition and disposal of buildings.

2.0.4 Passive techniques

Non-mechanical techniques consuming no or less energy, directly using the natural conditions of sunshine, wind, air temperature, humidity, terrain, plants, etc. on site through optimizing the planning and building design to reduce the heating, air-conditioning and lighting loads of buildings, so as to raise the indoor and outdoor environmental performance.

2.0.5 Active techniques

Energy-consumed mechanical techniques employed in order to raise the indoor comfort and realize indoor and outdoor environmental performance.

2.0.6 Incremental cost of green building

Increase or decrease in the initial investment cost due to the implementation of the green building idea and strategy.

3 Basic Requirements

3.0.1 Scientific outlook on development shall be used as a guide for developing urban-rural planning, and the associated indexes of low-carbon ecological planning shall be determined at the detailed planning stage in combination with the land use situation, so as to guide the green design of civil buildings at the subsequent stage.

3.0.2 Green design of civil buildings shall, in combination with the specific situation of the project, enforce the planning indexes made at the planning stage and associated building indexes, and realize the expected green building targets.

3.0.3 Design shall comply with the principle of adjusting measures to local conditions, and combine the climate, resource, eco-environment, economy, humanity and other features of Beijing.

3.0.4 Design shall synthesize the technical and economic characteristics of the building life cycle, and use the planning design patterns, architectural forms, techniques, materials and equipments in favor of promoting sustainable development.

3.0.5 Design shall embody the ideas of sharing, balance and integration. In the course of design, planning; architecture; structure; water supply and sewerage; heating, ventilation, air-conditioning and cooling; gas; electricity and intelligentization; indoor design; landscape; economy; and other specialties shall work cooperatively.

4 Index System

4.1 General Requirements

4.1.1 The low-carbon and ecological planning at the detailed planning stage shall be represented and controlled by the key indexes of land use planning, traffic planning, resource utilization and eco-environment. The green design at the building design stage shall be represented and controlled by the key indexes of architecture; structure; water supply and sewerage; heating, ventilation, air-conditioning and cooling; electricity; landscape; and interior decoration.

4.1.2 The calculation method, value and scope of application of each index shall be in accordance with the requirements specified in Table 4.2.2 and Table 4.3.2 in this chapter.

4.2 Low Carbon Ecological Design index System at Detailed Planning Stage

4.2.1 The indexes in Table 4.2.2 shall be made for the low-carbon and ecological planning at the detailed planning stage.

4.2.2 The key indexes of the low-carbon and ecological planning at the detailed planning stage shall be in accordance with the requirements specified in Table 4.2.2.

Table 4.2.2 Low Carbon Ecological Design index Table at Detailed Planning Stage

No.	Classification	Content	Definition and Calculation Method	Recommended Value	Remarks
P1	Land Use Planning	Plot scale	The length and width range of the plot enclosed by branch roads	150m~250m	Mainly applicable to new cities, center cities and old cities refer to this standard
P2		Residential land area per capita	1) The ratio of residential land area to the accommodated resident population in it. 2) The population in the residential area is calculated according to the ratio of 2.8 persons/household	1st~3rd floor $\leqslant$49m^2/capita; 4th~6th floor $\leqslant$32m^2/capita; 7th~9th floor $\leqslant$27m^2/capita; 10th floor and above $\leqslant$17m^2/capita	Applicable to the residential projects of new cities, center cities and old cities
P3		Underground building floor area ratio	The ratio of total underground floor area to site area of the plot	High-rise building $\geqslant$0.5; Multi-storey building $\geqslant$0.3	Applicable to various projects of new cities, center cities and old cities, excluding buildings without underground space utilization conditions due to the impact of geological conditions, foundation forms and municipal infrastructures, etc

continue 4.2.2

No.	Classification	Content	Definition and Calculation Method	Recommended Value	Remarks
P4	Land Use Planning	Accessibility to public facilities	1) It refers to meeting the walking distance from the building entrance/exit to above 6 types of community public service facilities 2) Community public service facilities mainly include kindergartens, primary schools, community health service stations, culture activity stations, small community commerce, post offices, bank, community management and service centers, indoor and outdoor sports fitness facilities, etc	≤500m	Applicable to the residential projects of new cities, center cities and old cities
P5		Accessibility to urban open space	1) The walking distance from the building main entrance/exit in the planning area to the surrounding urban open space (excluding community-level parks). 2) The urban open space refers to the ground or water area covered with no or few buildings or structures in cities, including urban public green spaces (excluding road attached green spaces such as green belts and street trees), parks, squares, etc	≤500m	Applicable to various projects of new cities, center cities and old cities
P6		Ratio of the number of jobs within 1km of rail transit stations to daily everage passenger flow volume	The ratio of the number of available jobs within 1km of the rail transit station to daily average unidirectional passenger flow volume designed for the station	≥10%	Applicable to various projects of new cities, center cities and old cities
P7		Ratio of barrier-free residences (guest rooms)	1) The ratio of the number of barrier-free residences (guest rooms) meeting barrier-free residence design standard to the total residences (guest rooms) in the project 2) The barrier-free residence refers to a residence which has the entrance/exit, passageway, communication, furniture, kitchen, bathroom, etc. provided with barrier-free facilities, and the spatial scale convenient for the mobility impaired to move safely. 3) The barrier-free guest room refers to a guest room which has the entrance/exit, passageway, communication, furniture, kitchen, bathroom, etc. provided with barrier-free facilities, and the spatial scale convenient for the mobility impaired to move safely	Residential area ≥2%; Hotel ≥1%	Applicable to the residential projects of new cities, center cities and old cities

continue 4.2.2

No.	Classification	Content	Definition and Calculation Method	Recommended Value	Remarks
P8	Traffic Planning	Coverage rate of public transport stations	The ratio of land area with a walking distance of less than 500m from the main entrance/exit of the main functional building to the nearest public transport station to total land area	100%	Applicable to various projects of new cities, center cities and old cities
P9		Outdoor parking ratio	The ratio of outdoor parking spaces to total parking volume of the project	Residential building ≤10%; Luxury apartment and villa ≤7.5%. The requirements of public buildings are presented according to the project and site features	Applicable to various projects of new cities, center cities and old cities
P10	Resource Utilization	Energy consumption per unit floor area	1) to Public buildings: the indicator Refers to the energy consumption (excluding urban municipal heating) due to various activities in buildings, including that of air conditioning, lighting, socket outlets, elevators, cooking, various service facilities and special functional equipment, in kWh/($m^2 \cdot a$). 2) to Residential buildings: the indicator Refers to the indicator of heat loss of building. The heat consumed on the unit floor area in unit time and supplied by indoor heating equipment under the condition of calculating the outdoor average temperature in the heating period, in order to keep the indoor design calculation temperature, in W/m^2.	Public buildings: Large administrative office ≤74kWh/($m^2 \cdot a$); Large business office ≤135kWh/($m^2 \cdot a$); General office (split air conditioner) ≤37 kWh/($m^2 \cdot a$); Large shopping mall and supermarket ≤137 kWh/($m^2 \cdot a$); General shopping mall and supermarket ≤75kWh/($m^2 \cdot a$); Large hotel ≤160kWh/($m^2 \cdot a$); General hotel ≤80 kWh/($m^2 \cdot a$); Large education ≤90 kWh/($m^2 \cdot a$); General education ≤22kWh/($m^2 \cdot a$); Medical ≤138 kWh/($m^2 \cdot a$) Residential buildings: 3rd floor and below ≤14.5W/m^2; 4th floor ~ 8th floor ≤10.5W/m^2; 9th floor ~ 13th floor ≤9.5W/m^2; 14th floor and above ≤8.5W/m^2	Applicable to various projects of new cities, center cities and old cities. "Large" refers to the building with a floor area of above 20,000 m^2, and see Table B.0.9 in Appendix B of this standard for the data source

continue 4.2.2

No.	Classification	Content	Definition and Calculation Method	Recommended Value	Remarks
P11	Resource Utilization	Contribution rate of renewable energy	The ratio of conventional energy consumption saved by renewable energy in the project in whole year to total energy consumption in the project in whole year Contribution rate of renewable energy $$= \frac{\text{Saving by renewable energy in the project (tce)}}{\text{Total energy consumption in the project (tce)}} \times 100\% \quad (4.2.2\text{-}1)$$ $$= \frac{\text{Conventional energy consumption before using renewable energy (tce)} - \text{Conventional energy consumption after using renewable energy (tce)}}{\text{Total energy consumption in the project (tce)}} \times 100\% \quad (4.2.2\text{-}2)$$ The renewable energy includes non-fossil energy such as solar energy, geothermal energy, biomass energy and wind energy	Residential buildings $\geqslant 6\%$; Office buildings $\geqslant 2\%$; Hotel buildings $\geqslant 10\%$	Applicable to the residential, office and hotel projects of new cities and center cities, and the similar projects of old cities refer to this standard
P12		Average daily rated water consumption	Average daily water consumption index for the project	The average daily water consumption for residential buildings is $\leqslant 110\text{L}/(\text{capita} \cdot \text{d})$, and the water consumption for other buildings is the median according to the requirement of the standard GB 50555 *Standard for Water Savig Design in Civil Building*	Applicable to various projects of new cities, center cities and old cities
P13		Classified collection rate of domestic waste	The percentage of domestic waste amount realizing classified collection in total domestic waste output in the area, or the ratio of inhabitants implementing separate waste collection to total inhabitants in the target area	$\geqslant 90\%$	Applicable to various projects of new cities, center cities and old cities

continue 4.2.2

No.	Classific-ation	Content	Definition and Calculation Method	Recommended Value	Remarks
P14		Rainwater runoff discharge	The runoff of rainwater discharged to urban municipal rainwater pipe network or natural water in the site	The total rainwater discharge in the site after development is equal to or less than that before development	Applicable to various projects of new cities, center cities and old cities
P15	Eco-environment	Ratio of sunken green space	1) The percentage of sunken green space area in total green space area (excluding the green spaces above the underground space with an overburden soil layer of less than 1.5m) in the site. 2) The sunken green space refers to the green space 5cm ~ 10cm lower than the surrounding road or ground surface. The construction of sunken green space also includes tree pools, rain gardens, grass swales, dry ponds, wet ponds, etc	≥50%	Applicable to various projects of new cities, center cities and old cities
P16		Permeable pavement rate	1) The percentage of permeable pavement area in rigid pavement area (including various roads, squares, parking lot, excluding the fire fighting access and ground above the underground space with an overburden soil layer of less than 1.5m) in this area. 2) The permeable pavement shall meet the associated requirements of the product standard JC/T 945 *Water Permeable Brick*. The hollowed-out pavement with a hollowed-out rate equal to or greater than 40% will not be included into the permeable pavement or rigid pavement. 3) The base course of the permeable pavement shall meet the associated requirements of GB 50400 *Engineering Technical Code for Rain Utilization in Building and Sub-district* and DB11/T 686 *Specification for Construction and Acceptance of Water Permeable Brick Pavement*	≥70%	Applicable to various projects of new cities, center cities and old cities

continue 4.2.2

No.	Classification	Content	Definition and Calculation Method	Recommended Value	Remarks
P17	Eco-environment	Green Ratio	1) The ratio of the sum of various green spaces within the boundary line to the site area (%), which shall be calculated according to the following formula: Green Ratio $= \dfrac{\text{Area of various green spaces within the boundary line (km}^2)}{\text{Site area whithin the boundary line (km}^2)} \times 100\%$ (4.2.2-3) 2) Green spaces shall include: concentrated green spaces, green spaces between houses, green spaces attached to public facilities, etc. within the boundary line, which include the roof greening of underground or semi-underground buildings meeting the tree planting and greening overburden soil requirement in Beijing, but exclude artificial green spaces on roofs and terraces	Residential areas of newurban area and downtown area $\geqslant$ 30%, residential areas of old city area $\geqslant$ 25%, the requirements of public buildings are presented according to the project and site features	Applicable to various projects of new cities, center cities and old cities
P18		Roof greening ratio	1) The ratio of greened roof area to greenable roof area. 2) Roof greening shall be implemented for newly-built and rebuilt buildings (including skirt buildings), having non-pitched roofs, with fewer than 12 storeys and lower than 40m height. 3) Roofs such as with a pitch of greater than 15°, long-span lightweight roofs, roofs provided with outdoor equipment, etc. can not be deemed as greenable roofs	$\geqslant$30%	Applicable to public building projects of new cities, center cities and old cities
P19		Woodlot ratio	1) The ratio of woodlot area to greening land area within the boundary line. 2) Woodlots refer to the lands with trees planted within urban public green spaces, green buffers and other construction lands, and the woodlot area is calculated according to the vertical projection of tree crowns. The distance of trunks between two adjacent trees shall be less than 10m	Woodlot ratio in public green spaces $\geqslant$ 25%; woodlot ratio in green buffers $\geqslant$60%; and woodlot ratio in other construction lands $\geqslant$40%	Applicable to various projects of new cities, center cities and old cities refer to this standard

continue 4.2.2

No.	Classification	Content	Definition and Calculation Method	Recommended Value	Remarks
P20	Eco-environment	Native plant index	1) The ratio of native plant species to all plant species in the planning area. 2) The native plant indicator shall be calculated according to the following formula: $$P_3 = \frac{N_{b3}}{N_3} \quad (4.2.2\text{-}4)$$ Where: P_3——Native plant index; N_{b3}——Total native plant species in the area; N_3——Total plant species in the area. 3) Native plants include: ① Wild plant species and their derived varieties naturally grown natively; ② Naturalized species (not originated natively but has escaped) and their derived varieties; ③ Domesticated species (not originated natively, but has grown natively for a long time and completing their life story) and their derived varieties, except plant species in living collection, germplasm resources and introduction experiment for scientific research. 4) No statistics were considered not satisfied the index	⩾0.7	Applicable to various projects of new cities, center cities and old cities

4.3 Green Design Index System at Building Design Stage

4.3.1 The indexes in Table 4.3.2 shall be made at the green building design stage.

4.3.2 The key indexes of the green design at the building design stage shall comply with the requirements of Table 4.3.2.

Table 4.3.2 Green Building Design Index Table

No.	Classification	Content	Definition and Calculation Method	Recommended Value	Remarks
D1	Architecture Specialty	Compliance ratio of barrier-free design	The ratio of the number of barrier-free facilities that meet the design requirement in the building to the total number of barrier-free facilities that are required by the GB 50763 *Codes for Accessibility Design* - at the building entrance, elevator, bathroom, etc.	100%	Applicable to residential buildings and public buildings
D2		Distance from building entrance/exit to public transport station	The shortest walking distance from building entrance/exit to public transport station	⩽500m	Applicable to residential buildings and public buildings

continue 4.3.2

No.	Classification	Content	Definition and Calculation Method	Recommended Value	Remarks
D3	Architecture Specialty	Energy-saving design indexes of building envelope	Including the indicators such as Shape factor, Window to wall ratio, roof transparent area ratio, openable area ratio of external window and heat transfer coefficient of building envelope	Meeting the requirements of *the Design Standard for Energy Efficiency of Residential Buildings* and *Design Standard for Energy Efficiency of Public Buildings* of Beijing	Applicable to residential buildings and public buildings
D4		Area ratio of active exterior-shading	The ratio of the area of external windows with active exterior-shading facilities in main spaces at west facade to the total area of external windows in that direction	Meeting the *Design Standard for Energy Efficiency of Residential Buildings* of Beijing	Applicable to residential buildings
D5		Cost ratio of purely decorative components	The ratio of the cost of decorative components without functional value to total cost of the project	Residential buildings $<2\%$; Public buildings $<5\textperthousand$	Applicable to residential buildings and public buildings
D6		Area ratio of space enclosed by recycled partitions	1) The ratio of the total area of rooms enclosed by recycled partitions (excluding office room area $\geq 100m^2$ and other rooms area $\geq 500m^2$) to the total area of indoor spaces with convertible functions. 2) Recycled partitions include plank partition, framework partitions, movable partitions, glass partitions, etc., and non-recycled partitions include non-load-bearing masonry partition, etc	$\geq 30\%$	Applicable to the buildings needing to re-divide space, such as office, shopping mall or meeting space
D7		Utilization rate of waste-reuse materials	1) The ratio of the weight of waste-reuse materials to that of similar building materials. 2) Material-reuse materials refers to the building materials produced by using waste as raw materials on the premise of ensuring performance, safety, health and environmental protection, the consumption of such materials shall be great and the mixing amount of waste shall be greater than 20%	The ratio of the consumption to similar building materials is $\geq 30\%$	Applicable to residential buildings and public buildings
D8		Utilization rate of recycled materials	1) The ratio of the weight of recycled materials to that of total building materials. 2) Recycled materials refer to the materials that can be recycled for many times by changing the physical form of the material that cannot be reused and generating another material	$\geq 10\%$	Applicable to residential buildings and public buildings

continue 4.3.2

No.	Classification	Content	Definition and Calculation Method	Recommended Value	Remarks
D9	Architecture Specialty	Indoor noise compliance ratio of main function space	The ratio of the number of function rooms, which Both the indoor noise and the sound-insulating standard of envelope components meet the requirement of the code GB 50118 *Code for Sound Insulation Design of Civil Buildings*, to the total number of building function rooms	100%	Applicable to residential buildings and public buildings
D10		Ratio of high strength steel bars	The ratio of the weight equivalent value of HRB400 stressed steel bars to the total weight equivalent value of stressed steel bars in reinforced concrete structure	To Structure of 6~9-storey buildings: ≥70%; to structure of 10-storey and above buildings: ≥80%.	Applicable to residential buildings and public buildings
D11		Ratio of high strength concrete	The ratio of the weight equivalent value of C50 concrete in the vertical load-bearing structure to the total weight equivalent value of concrete in the vertical load-bearing structure in the reinforced concrete structure of high-rise buildings above 60m	To Residential buildings: (number of storeys - 20)/number of storeys; To Public buildings: (number of storeys - 15)/number of storeys	Applicable to residential buildings and public buildings
D12		Ratio of high performance steel	The ratio of the weight equivalent value of high performance steel above Q345 to the total weight equivalent value of steel in high-rise steel buildings	≥70%	Applicable to residential buildings and public buildings
D13	Water Supply and Sewerage Specialty	Utilization rate of water-saving devices and equipments	The ratio of the number of water devices and equipments meeting the requirements of CJ 164 *Domestic Water Saving Devices*, GB/T 18870 *Technical Conditions for Water Saving Products and General Regulation for Management*, and DB11/ 343 *Technical Specification for Water Saving Apparatus* of Beijing to the total number of water devices and equipments in the building	100%	Applicable to residential buildings and public buildings
D14		Utilization rate of nontraditional water source	The percentage of the annual water consumption of nontraditional water source, such as recycled water and rainwater, instead of municipal water supply or ground water supply for landscape, greening, toilet flushing, etc. to the total annual water consumption	To Residential buildings ≥10%; To Office buildings and shopping malls ≥20%; To Hotels ≥15%	Applicable to residential buildings and public buildings
D15		Utilization rate of water-saving irrigation for green spaces	The ratio of green space area using water-saving irrigation system to the sum of the green space area. The water-saving irrigation system includes sprinkling irrigation system, micro-irrigation system, drip irrigation system, etc	100%	Applicable to residential buildings and public buildings

continue 4.3.2

No.	Classification	Content	Definition and Calculation Method	Recommended Value	Remarks
D16	Heating, Ventilation, Air-conditioning and Cooling Specialty Architecture Specialty	Summated refrigerating coefficient of performance (SCOP) of water chilling (heat pump) unit with centralized cooling source	Where: Qc - Cooling capacity output from cooling source under nominal conditions (kW) Ee - Electricity consumption input to cooling source under nominal conditions (kW); for centrifugal machines, screw machine and pistons, Ee includes the electricity consumption of refrigerators, cooling pumps and cooling towers; for water-source and ground-source heat pumps, Ee includes the electricity consumption of refrigerators, cooling pumps, and water pumps for taking ground water and recharging water	Higher than the requirements of the *Design Standard for Energy Efficiency of Public Buildings* of Beijing.	Applicable to residential buildings and public buildings
D17		Coefficient of performance (COP) of water chilling (heat pump) unit with centralized cooling source	The ratio of actual refrigerating capacity to actual input power when the water chilling (heat pump) unit on the refrigeration of centralized cooling source under rated and specified conditions	At least One level higher than the requirements of the *Design Standard for Energy Efficiency of Public Buildings* of Beijing	Applicable to residential buildings and public buildings
D18		System transfer and distribution efficiency	Including the requirements for parameters such as the design value of the ratio of electricity consumption to transferied heat quantity of heating hot water circulating pump, the ratio of electricity consumption to transferied heat quantity of air-conditioning hot water circulating pump, the ratio of electricity consumption to transferied cooling quantity of air-conditioning cold water circulating pump, and the power consumption of unit air volume of fan	Not lower than the requirements of the *Design Standard for Energy Efficiency of Residential Buildings* and *Design Standard for Energy Efficiency of Public Buildings* of Beijing	Applicable to residential buildings and public buildings
D19	Ectric Specialty	Lighting power density (W/m²)	The lighting installation power per unit area in building rooms or places, including the installation power of light sources, ballasts or transformers	Not higher than the target value of the current national standard GB 50034 *Standard for Lighting Design of Buildings*	Applicable to residential buildings and public buildings
D20		Target energy efficiency of transformer	The maximum standard value of allowable no-load loss and load loss for power transformers under standard specified test conditions	Meeting the requirements of the evaluating values of energy conservation of the Minimum Allowable Value of GB 20052 *Energy Efficiency and the Evaluating Values of Energy Conservation for Three-phase Distribution Transformers*	Applicable to residential buildings and public buildings

continue 4.3.2

No.	Classification	Content	Definition and Calculation Method	Recommended Value	Remarks
D21	Landscape Design	Lighting power density of building facade nightscape	The lighting installation power per unit area for the nightscape lighting of building elevations, including the installation power of light sources, ballasts or transformers	Meeting the requirements of JGJ/T 163 *Code for Lighting Design of Urban Nightscape*	Applicable to residential buildings and public buildings
D22		Solar radiation absorption factor of rigid pavement	The ratio of solar irradiance absorbed by rigid pavement to total solar irradiation that project on it	0.3~0.7	Applicable to residential buildings and public buildings
D23		Shading ratio of outdoor parking spaces	The ratio of the outdoor parking space area shadowed by the vertical projections of tree crowns, shading facilities, etc. to the total area of the outdoor parking spaces	≥30%	Applicable to residential buildings and public buildings
D24		Shading ratio of sidewalk and bicycle road	The ratio of the road length shadowed by trees to total road length within the boundary line	≥75%	Applicable to residential buildings and public buildings
D25		Number of arbors/100m² green space	The average number of arbors/100m² outdoor green spaces	≥3pcs	Applicable to residential buildings and public buildings
D26		species of woody plants	1) This indicator refers to the woody plant species within the planning area. 2) Woody plants refer to the plants with flourishing xylem and hard texture in the stem, generally upright, having a long life, grown for many years, corresponding to herbaceous plants. Woody plants can be divided into arbors and shrubs according to the form	Not less than 30 species for the project land area less than 50,000m²; Not less than 35 species for the project land area between 50,000m²~100,000m²; Not less than 40 species for the project land area larger than 100,000m²	Applicable to residential buildings and public buildings
D27	Interior Decoration	Ratio of construction and decoration integration	The ratio of the floor area of residential buildings realizing construction and decoration integration to total floor area of residential buildings in the project	100%	Applicable to residential buildings only

5 Scheme and Design Documents of Green Building

5.1 Scheme of Green Building

5.1.1 At the scheme stage of building project, the scheme of green building should be made, and the scheme proposal of green building should be compiled.

5.1.2 The scheme of green building aims at defining the project orientation of green building, indicators of green building, corresponding technical strategies, and cost-benefit analysis.

5.1.3 The scheme of green building shall include the following contents:

1 Early survey;
2 Project orientation and target analysis;
3 Scheme of green design concept and analysis of implementation strategy;
4 Technical and economic feasibility analysis.

5.1.4 Early survey should include site analysis, market analysis and social environment analysis, and should meet the following requirements:

1 The site analysis should include the geographic location of project, eco-environment of site, climatic environment of site, terrain and landform, surroundings around site, road traffic, conditions for municipal infrastructure planning, etc.;

2 The market analysis should include the functional requirements, market demand, usage pattern, technical conditions, etc.;

3 The social environment analysis should include regional resources, cultural environment and quality of life, regional economic level and development space, opinions and advice from the surrounding public, incentive policies for green building in the located region, etc.

5.1.5 The project orientation and target analysis should include the following contents:

1 Analyze the own characteristics and requirements of the project;

2 Reach the corresponding level required by the national standard GB/T 50378 *Evaluation Standard for Green Building* or the local standard DB 11/T825 *Evaluation Standard for Green Building* of Beijing;

3 Determine the appropriate roadmap for implementation, and meet the corresponding indicator requirements.

5.1.6 The scheme of green design concept and analysis of implementation strategy should meet the following requirements:

1 Comply with the principles of giving priority to passive techniques and optimizing active techniques, and reasonably select the applicable technology;

2 Select integrated technology;

3 Select high-efficiency building products and equipment as well as green and environment-friendly building materials;

4 If the target of green building can't be satisfied by the existing conditions, employ the techniques of adjustment, balance and compensation.

5.1.7 The technical and economic feasibility analysis should include the following contents:

  1 Technical feasibility analysis;
  2 Economic feasibility analysis;
  3 Benefit analysis;
  4 Risk analysis.

5.2 Organization of Green Building Design

5.2.1 The project builder shall organize a green building team. The team members shall include the units related to the project construction, usage and operating management, such as the units of builder, design, consultation, construction, supervision, property management, etc.

5.2.2 The consultation unit of green building should develop work at the scheme design stage or earlier, and shall fully respect the integral requirements of the project and provide the scheme of green building, advice for design, and technical support.

5.2.3 The design unit of green building shall reasonably deploy professional technicians, and should appoint a director for green building design, every specialty shall cooperatively work at each stage according to the principle of technical sharing, balance and integration.

5.2.4 In the design documents all the relevant specialties shall give clear design requirements for green building to the special design, and shall check and confirm whether the special design meets the corresponding requirements or not.

5.3 Design Documents of Green Building Design

5.3.1 The proposal for project shall be drafted in accordance with the requirements for regional low-carbon and ecological planning. The special subject of green building shall be set to make the target for the green building to be reached, and the incremental cost of green building shall be listed in the investment estimation.

5.3.2 The feasibility study report of project shall be drafted in accordance with the requirements for regional low-carbon and ecological planning. The special subject of green building shall be set to make overall analytic demonstration for the requirements given in this standard and define the implementation strategies for green building.

5.3.3 The draft unit of detailed planning shall draft the planning according to the contents in Chapter 6 and the requirements in Section 4.2 "Low Carbon Ecological Design Index System at Detailed Planning Stage", and the design documents of the planning shall reflect the corresponding contents.

5.3.4 The tender documents of project scheme design shall set the special subject of green building in the design document according to the design requirements in the bid document for design.

5.3.5 The scheme design document shall set the special subject of green building, including the target for green building of project, green building means and technologies designed to use, investment estimation, etc., and the Integrated Review Sheet shall be completed according to the format in Appendix A of this standard.

5.3.6 The instructions of primary design shall set the special subject of green building, and the Integrated Review Sheet shall be completed according to the format in Appendix A of this standard.

5.3.7 The instructions of construction drawing design shall set the special subject of green building. The special subject shall be set in the design instruction on architecture specialty, and the technical requirements related to design in the construction of green buildings and the operating management of building should be indicated. The Integrated Review Sheet shall be completed according to the format in Appendix A of this standard.

6 Planning and Design

6.1 General Requirements

6.1.1 The planning and design mentioned in this chapter include detailed planning and site plan.

6.1.2 The low-carbon and ecological design at the detailed planning stage shall take the comprehensive contents into consideration, such as space, transportation, energy, resources, environment, etc.

6.1.3 The site plan shall target at improving the outdoor environmental quality and ecological benefits, optimize the layout of architectural planning, and make eco-environmental restoration and ecological compensation for the site.

6.2 Land Use Planning

6.2.1 The site selection shall be in accordance with the following requirements:

1 The site selection shall meet the relevant requirements of ecologically restricted planning such as Beijing planning for limited construction area, etc., and shall complete the evaluation on the ecological applicability. No site shall be selected for construction in any ecological sensitive area, and no basic farmland shall be occupied;

2 The conditions of engineering geology, hydrogeology, seismic hazard and geological hazard shall be evaluated according to the regional safety, and it is forbidden to arrange any site selection in the radius of various hazards. The site selection shall be located out of the safety influential range of the danger of electromagnetic radiation, the danger of dangerous chemicals, and the hazards of pollutants and poisonous substances, etc.; simultaneously, the site selection shall ensure that the impact on the surrounding can comply with the requirements for the safety evaluation of environment;

3 Land selection shall give priority to renewable reconstruction land or waste land, and the reconstruction and utilization of industrial land shall meet the requirements for the safety evaluation of environment;

4 Site selection and construction should be performed around rail transit stations; the ratio of the number of available jobs within 1km of rail transit stations to daily average unidirectional passenger flow volume designed for the stations shall not be less than 10%;

5 Regions with favorable infrastructure conditions should be selected, and the building capacity shall be rechecked according to the infrastructure conditions.

6.2.2 Land use planning shall be in accordance with the following requirements:

1 Compact layout shall be available around rail transit stations; the integrated design shall be performed for land use, traffic and underground spaces around pivot rail transit stations;

2 The plot scale planning shall give consideration to the demand for convenience for walking out. For newly built districts in urban area, the plot enclosed by branch roads should not be larger than 150m~250m; for reconstruction of old areas, the reasonable plot scale shall be consummated by employing the techniques of denser road network, opening up road microcirculation;

3 It shall meet the requirements of traditional culture for sustainable development, and the spatial planning shall be matched with the urban featured context and texture;

4 Urban comprehensive public service centers shall be arranged around rail transit stations;

5 The old buildings which can be kept in the site should be reutilized, and new use functions shall be provided for the building in combination with the requirements for urban development;

6 Vertical design shall be reasonably performed to realize the economic balance of earthwork.

6.2.3 The layout of land functions shall be in accordance with the following requirements:

1 The land functions shall comply with the principle of balanced development for work and living, and the ratio of the number of available jobs in the land range or within the 1km around it to the total number of households in the same area should be controlled within the range of 0.6~1.6;

2 The land use planning shall take the moderate mixed utilization of functions into consideration, and the reasonable mixing configuration is required according to the space and functions of locations.

6.2.4 The planning of public facilities shall be in accordance with the following requirements:

1 Public service facilities should be configured according to the community scale; aiming at the social problems of aging, vulnerable groups, difficulties in parking, etc., the public facilities for pension, helping the disabled, accessibility, parking, etc. shall be moderately increased;

2 Public facilities shall be reasonably arranged, the accessibility to six or more of the community public service facilities should not exceed 500m, including kindergartens, primary schools, community health service stations, culture activity stations, small community commerce, post offices, bank outlets, community management and service centers, indoor and outdoor sports fitness facilities, etc. ;

3 The ratio of accessible housing in residential districts shall be higher than 2%, and the ratio of accessible guest rooms in hotels shall be higher than 1%;

4 The functions of business, retails, etc. should be arranged around public transport stations; on both sides of the street where public buildings are in centralized layout, the favorable walking spaces may be created by controlling the reasonable ratio of buildings along routes, and the ratio of buildings along routes should be higher than 50%.

6.2.5 The control of the construction strength of land shall be in accordance with the following requirements:

1 The service strength of land around rail transit stations shall meet the requirements of the standard for land saving of urban construction of Beijing;

2 Underground spaces shall be reasonably planned and designed to increase the comprehensive utilization efficiency of land. The underground building floor area ratio should not be less than 0.3 for multistory buildings and not be less than 0.5 for high-rise buildings;

3 The land for intensive utilization shall be saved, and the residential land area per capita shall meet the value requirements in Beijing's evaluation standard for green building, that is, it should not be larger than 49m^2/capita for Floor 1~3, not be larger than 32m^2/capita for Floor 4~6, not be larger than 27m^2/capita for Floor 7~9, and not larger than 17m^2/capita for Floor 10 and above (including Floor 10).

6.3 Transportation Planning

6.3.1 The planning for road and public transport system shall be in accordance with the following requirements:

    1 It is a priority to develop the public transport and optimize the public transport route network, and the coverage rate of public transport stations shall meet the requirement that the walking distance between the main entrance/exit of building and the public transport stations is less than 500m;

    2 The road network density and road area shall be reasonably determined, and the layout of urban branch road system and the density of branch road network shall be reasonably planned;

    3 Sidewalks and bicycle roads conveniently communicated with the surrounding public facilities and public transport stations should be provided at the entrance/exit of land for convenience of going out on foot and by bicycle;

    4 In old urban districts, the road network planning shall give comprehensive consideration to the conditions of original ground and underground building, municipal conditions and the characteristics of the original roads. The streets of historical values shall be kept and utilized.

6.3.2 For new district planning and the existing district conditional for reconstruction, the planning for walking and bicycle systems shall be in accordance with the following requirements:

    1 The bicycle road network is composed of the bicycle roads on both sides of urban road, lanes, community roads and special bicycle roads, and it shall ensure the continuous riding of bicycles. Sidewalks, pedestrian roads, pedestrian facilities for getting across streets, etc. shall form the integral urban walking system together with the walking systems of residential districts, business areas, stations and urban squares;

    2 The width of bicycle roads on both sides of urban roads shall be 3m~4m for pavements of express way and arterial roads, 2m~3m (it should be 3m under normal conditions) for secondary main roads, and 2m for branch roads. It may be appropriately widened for the roads where the bicycle flow exceeds 3,000 units in rush hours. The width of bicycle road may be 1m for the branch roads with the boundary line width of 15m. The width of bicycle road at crossings shall not be less than that on the road section;

    3 For urban secondary main roads or those of higher grade, physical separation must be arranged between motor vehicle roads and bicycle roads. For urban branch roads with bigger traffic volume, separation facilities shall also be provided between motor vehicle roads and bicycle roads according to the conditions;

    4 For residential districts, public service facilities, public transport stations and public transport pivots, enough and convenient parking facilities for bicycles shall be arranged nearby, and the overall scale of parking facilities for bicycles in transit station may be estimated according to 10%~15% of the total number of passengers in the station (the upper limit should be required for stations outside the Third Ring Road);

    5 Sidewalks shall be set in the boundary lines of urban roads, and crosswalks shall be set at all crossings. Convenient walking systems communicated with main public transport stations, public service centers and surrounding functional zones should be arranged within the boundary lines of urban roads, and the walking systems in plots may be designed in combination with breeze channels. The walking traffic facilities shall meet the requirements for accessibility;

6 The walking and bicycle systems shall be designed in combination with greening design and landscape design, supporting rest facilities shall be provided, and greening and shading techniques shall be employed to improve the comfort of sidewalks and bicycle roads. The shading ratio of sidewalk and bicycle road should not be less than 75%.

6.3.3 The planning of static traffic system shall be in accordance with the following requirements:

1 The parking of motor vehicles shall meet the requirement for land saving, and underground parking and three-dimensional parking shall be selected in priority;

2 The number of parking spaces for motor vehicles shall be reasonably determined to control the ratio of open parking spaces for motor vehicles. The ratio of the number of open parking spaces to the total parking volume shall not exceed 10% for residential districts and shall not exceed 7.5% for high-grade apartments;

3 The parking lot (garage) for motor vehicles shall contain a certain proportion of parking spaces for electro cars and indexes for charging stations;

4 The parking place for bicycles shall be reasonably arranged, and the parking place for bicycles should not exceed 150m from entrance/exit of buildings. Parking facilities for bicycles shall be arranged around rail transit stations and public transport stations;

5 The ecological design shall be considered for parking lots, and plants or sunshades shall be used for increasing the shading ratio of open parking spaces. The compliance ratio of greening parking shall be completely satisfied.

6.4 Resource Utilization

6.4.1 Energy utilization shall be in accordance with the following requirements:

1 Integral energy planning shall be performed for regions and buildings. Passive techniques shall beemployed in priority to reduce the energy consumption, and the unit building area of various buildings shall meet the requirements for design indexes in this standard. The planning shall improve the contribution rate of renewable energy, and the contribution rate of renewable energy should not be less than 6% for residential buildings, not be less than 2% for office buildings, and not be less than 10% for hotels and restaurants;

2 Solar energy shall be utilized in priority, and the regional solar resources shall be investigated and evaluated to determine the reasonable manner of utilization;

3 When geothermal energy is utilized, the delamination of underground soil, moisture distribution and infiltration capacity shall be investigated to evaluate the impact of geothermal energy exploitation on the underground spaces nearby, subterranean animals and plants or eco-environment, and techniques shall beemployed to prevent pollution to soil and underground water;

4 The distributive energy should be reasonably utilized according to the municipal conditions.

6.4.2 The utilization of water resources shall be in accordance with the following requirements:

1 The conditions of regional water resources shall be surveyed in details, and the water resources shall be reasonably planned and utilized in combination with the special planning for urban water environment;

2 The water quantity and water quality of regional water shall be estimated and evaluated according to the use requirement, so as to reasonably determine the water saving quota and water distribution scheme;

3 The applicable water treatment technology and facilities shall be usedfor reinforcing the recycling of water resources and improving the utilization ratio of urban reclaimed water resources.

6.4.3 The planning of wastes shall be in accordance with the following requirements:

1 The reducing principle for domestic solid wastes shall be followed to reasonably arrange the separate collection facilities of wastes, and the separate collection rate of domestic waste shall be higher than 90%;

2 The waste recovery and reuse system shall be arranged, and the closed refuse transfer station of residential districts shall have the function of separate collection of wastes, whereas communities shall be matched with recovery sites for renewable resources, and it may be considered to be arranged in combination with the closed refuse transfer station.

6.5 Ecological Environment

6.5.1 The planning of eco-environment shall be in accordance with the following requirements:

1 The ecological balance, biodiversity and connectivity of regional ecological system shall be kept for the land and the surrounding;

2 Measures shall be taken to develop ecological compensation and restoration while development and construction;

3 The site design shall be adapted to the original terrain and landform to protect and improve the ecological value of land, and the building layout in the site shall be organically combined with the existing trees;

4 Greening rate shall be reasonably determined, and it shall be no less than 30% for new and central urban residential districts, and no less than 25% for old urban residential districts. The open space systems of urban public green spaces, parks, squares, etc. shall be reasonably planned, and the accessibility to open space should not exceed 500m;

5 The roof greening ratio shall be reasonably determined, and the greening ratio for the greenable roof area shall not be less than 30%;

6 The ecological effect of green space system shall be ensured to reasonably determine the woodlot ratio, and the woodlot ratio should not be less than 25% for public green spaces, not be less than 60% for protective green spaces and not be less than 40% for other lands for construction; meanwhile, the native plant index shall be improved and shall not be less than 0.7.

6.5.2 The water environment design for the planned construction land shall be in accordance with the following requirements:

1 The planning and design schemes for water environment shall be reasonably determined, so as to ensure the integrity, ecological features, operability and sustainable comprehensive utilization, and improve the economic and environmental benefits;

2 Wet lands and surface water bodies shall be protected, and it is forbidden to destroy any regional water system in order to maintain the water quantity and water quality of surface water;

3 The paths for rainwater runoff in the site shall be reasonably planned, the techniques of rainwater infiltration, regulation and storage shall be employed so that the total rainwater discharge in the site after development will not be larger than that before development, and the rainwater system and rainwater environment should be optimized by using the auxiliary technology of simulation;

    4 The rainwater infiltration facilities of sunk green spaces, rain gardens, grass swales, etc. shall be reasonably designed in combination with the specific conditions, so as to supplement and cultivate the underground water resources and create a favorable hydrological eco-environment, and the ratio of sunk green space should not be lower than 50%.

6.5.3 The wind environment design for planned construction land shall be in accordance with the following requirements:

    1 The space layout of various outdoor activity sites shall be optimized according to the conditions and demands for outdoor wind environment, and the regional or land breeze channels shall be reasonably arranged to improve the wind environment.

    2 The building layout and shape shall create a favorable wind environment, so as to ensure the comfortable space for outdoor activity and favorable conditions for indoor national ventilation, and reduce the adverse impact of air flow on the regional microenvironment and the building itself. The building layout should avoid the unfavorable wind direction in winter. The width of multistory residential buildings should not exceed 60m, and the width of high-rise residential buildings should not exceed 50m;

    3 For the pedestrian area around buildings, the wind speed at 1.5m high shall not be higher than 5m/s, or the magnification coefficient of wind speed shall not be higher than 2. Computer simulation or wind tunnel test shall be employed for analyzing wind environment, and the simulation of building wind environment shall be set in accordance with the those specified in C.0.1of Appendix C.

6.5.4 The acoustical environment design for planned construction land shall be in accordance with the following requirements:

    1 Noise-sensitive Buildings shall be kept far away from noise sources; for fixed noise sources, appropriate techniques for sound isolation, noise reduction and shock insulation shall be employed; for the noise of main roads, the techniques of sound barrier or noise reduction pavements, etc. shall be employed;

    2 The active design shall be emphasized for acoustical environment, and scientific and technological means shall be applied to create the healthy and comfortable acoustical environment;

    3 The acoustical environment design forthe site shall meet the requirements of the current national standard GB 3096 *Environmental Quality Standards for Noise*. The current situation of noise around the land should be detected, the environmental noise after project implementation shall be predicted, and the simulation of noise environment shall be set in accordance with the those specified in C.0.5 of Appendix C.

6.5.5 The light environment design for planned construction land shall be in accordance with the following requirements:

    1 The building orientation shall be reasonably arranged according to the terrain, and natural light shall be fully used for reducing the energy consumption of indoor artificial lighting for buildings. The simulation of light environment should be employed to optimize the planning layout of buildings, and the simulation of light environment shall be set in accordance with those specified in

C. 0. 3 of Appendix C;

 2 For residential buildings and public buildings, the insolation standard and insolation interval shall meet the relevant requirements of the general rules for planning and design of construction works in Beijing;

 3 The site and road lighting shall be reasonably designed, no direct light of outdoor lighting shall be incident to the external windows of surrounding residential buildings, and no direct light of site and road lighting shall be incident to the air. The limit value for glare of ground reflected light should comply with the requirements of relevant standards;

 4 Reasonable design and material selection are required for the external surfaces of building to effectively avoid light pollution.

6.5.6 The thermal environment design for planned construction land shall be in accordance with the following requirements:

 1 The land use and buildings shall be reasonably arranged, and the natural ventilation of corridors shall be effectively used for reducing the heat island effect outdoors;

 2 Three-dimensional greening and multilayer greening shall be used, and arbor shading technique should be employed for parking lots, sidewalks, squares, etc. ;

 3 Corresponding techniques shall be employed to ensure the regional daily average strength of heat island index not to be higher than 1.5℃. The computer simulation should be employed to optimize the design of outdoor thermal environment, and the simulation of outdoor heat island shall be set in accordance with those specified in C. 0. 6 of Appendix C;

 4 The land for outdoor activities and pavement materials shall be selected on the basis of meeting the functional requirement for land use, and permeable pavement materials and permeable pavement structure shall be selected, whereas the permeable pavement rate shall not be less than 70%.

7 Architectural Design

7.1 General Requirements

7.1.1 Architectural design shall follow the principle of "passive techniques first" to optimize the building shape and space layout, promote the natural lighting and ventilation, reasonably optimize the envelope performances of thermal preservation, heat insulation, sunshading, etc., reduce the load of the heating, air-conditioning and lighting system, and improve the indoor comfortable level.

7.1.2 Architectural design shall give overall considerations to the factors of sound, light, heat, etc. inside and outside the site and balance the interactions between the factors in accordance with the surrounding environment and site conditions, so as to determine the reasonable layout, orientation, form and space of buildings. The building orientation should be north-south or close to north-south.

7.1.3 Various technical measures of architectural design shall be reasonably selected in combination with the performance advantages and disadvantages, applicable conditions, implementation effect, economic benefits, etc. after comprehensive comparative analysis.

7.1.4 The green building technologies such as sunshade members, light guiding members, air guiding members, solar thermal collectors, photovoltaic components, etc. shall be designed in integration with buildings.

7.1.5 The building form shall meet the requirements of building functions and technology. the structure and construction shall be reasonable.

7.1.6 At the stage of architectural scheme design, building performance and environment analytical technologies, for example, computer simulation should be used for analyzing and optimizing the orientation, direction, shape, envelope, interior space layout, etc., and they should be consummated and inspected in the process of further design.

7.2 Building Space Layout

7.2.1 Architectural design shall increase the utilization efficiency of space, and the space and facilities in buildings such as rest and communication space, conference facilities, should be shared.

7.2.2 Architectural design should be in favor of the socialized sharing of building space, and corridors, accessible roofs, etc. shall be used for providing externally shared public walkways, public activity space, public open space, exercise places, parking space, etc.

7.2.3 Architectural design should avoid unnecessary tall space and nonfunctional space, should avoid overlarge transitional and auxiliary space.

7.2.4 To consider fully the changes of building function, number of users and using method in the future, choose suitable width and story height, the indoor partitions shall improve the variability of the space function and the possibility of reconstruction.

7.2.5 The rooms where people will stay for long should be arranged in the positions with favorable sunlighting, daylighting, natural ventilation and vision.

7.2.6 The space requiring for privacy, such as residential bedrooms, hospital wards, guest rooms of hotels, etc. should avoid sight interference.

7.2.7 The space having the same or similar requirements for indoor thermal environment should be arranged together.

7.2.8 The rooms or places where people will live or work for long shall be far away from the room or place with noise, vibration, electromagnetic radiation and air pollution.

7.2.9 Mechanical rooms and piping shafts should be arranged close to the load center, and the route of pipelines shall be planned as a whole. The arrangement of mechanical rooms and piping shafts shall be convenient for maintaining, rebuilding and replacing of equipment and pipelines, and it shall be considered in design to reserve access doors, overhaul channels, space for expansion, channels for replacement, etc.

7.2.10 The stairs of daily use should be arranged near the main entrance and foyer of buildings, the stair cases should have natural ventilation and natural lighting arranged in combination with the fire-fighting escape stairs, and clear and noticeable indicative marks should be arranged at the entrances of the stair cases.

7.2.11 Buildings should have convenient parking space for bicycles and service facilities for bicycles. It is better to have showering and dressing facilities in qualified school and office buildings.

7.2.12 The positions of the building entrance shall be convenient for walkers and using public transit, and should be provided with pedestrian paths in convenient connection with public transport stations.

7.2.13 Fully use the space under shopping roof and other space which difficult to use.

7.2.14 The underground space shall be reasonably developed and used. Take measures to lead natural lighting and ventilation into underground space, the underground civil air defense facilities shall be fully used for the design of both peacetime and wartime use, and the frequently used underground space shall be provided perfect barrier-free facilities.

7.3 Building Envelope

7.3.1 The shape coefficient of buildings, the area ratio of window to wall, the thermodynamic engineering performance of envelope, the area of transparent part of roof, etc. shall meet the requirements of the current design standard for energy saving of Beijing. For conditioned buildings, the energy-saving standard for envelope may be raised, and the energy-saving rate of buildings may be calculated by means of weighing calculation or computer simulation analysis.

7.3.2 The wall design shall be in accordance with the following requirements:

1 The outer insulating layers in the positions where members extruded from the exterior wall and components attached to the wall should be closed so as to avoid the occurrence of thermal bridge;

2 Techniques for heat preservation and heat insulation shall be taken in the positions of reinforced concrete beams and slabs in the sandwich heat insulating exterior wall;

3 For buildings with continuous heating and air conditioning, the interior layer of the sandwich heat insulating exterior wall should be made of heavy and compact materials with good thermal inertia;

4 Insulating layers shall be provided for the partition walls and floor slabs between non-heat-

ing rooms and heating rooms;

5 Techniques for heat preservation and heat insulation should be taken between rooms with greater difference in temperature requirement or rooms having different periods of air conditioning and heating.

7.3.3 The design of external window shall be in accordance with the following requirements:

1 Don't use bay windows in the north. Shouldn't use large bay windows in other direction; the upper, lower and lateral non-transparent walls of bay windows shall be treated with heat insulation;

2 The gaps between the external window frame and the exterior wall shall be filled with high-efficient heat insulating materials and caulked with sealing materials;

3 The external window on the external thermal insulation wall should be arranged close to the outer side of the major part of the exterior wall, otherwise, the wall around the external window opening shall be treated with heat insulation;

4 Techniques of separating thermal bridge shall be taken for metal window frames and profiles of curtain wall.

7.3.4 The actual area of green roof should not be less than 30% of the area of greenable roof.

7.3.5 The conditional buildings should take techniques of heat insulation, such as light roof, ventilated roof, roof sunshade, etc.

7.3.6 For major use space, the external windows facing the east and west should taken adjustable sunshading, the external windows facing the south taken horizontal external sunshading, the scuttle should have sunshading facilities. Glass with better sunshading performance should be selected for windows facing the west and south. The area of the external window facing the west and provided with external sunshade facilities shall meet the requirements of the design standard for energy saving in Beijing.

7.3.7 For multistoried buildings, low-rise buildings and the low annex at the lower part of high-rise buildings, vertical greening should be provided for the exterior walls facing the west; for conditioned buildings, vertical greening should also be provided for the exterior walls facing the west, east and south. For the vines used for vertical greening of multistoried buildings, low-rise buildings and the low annex at the lower part of high-rise buildings, the horizontal planting length should not be less than 6% of the circumference of the building.

7.4 Building Material

7.4.1 Building materials shall be selected in accordance with the requirements about the relevant restrictive forbidden building materials and products promulgated by the nation and Beijing. Building materials currently generalized in Beijing, energetic materials and low carbon emission materials should be selected.

7.4.2 Building materials within 500km from the construction site should be selected.

7.4.3 Modularized and standardized design should be required for the planar and vertical dimensions, building components, etc. of residential buildings, hotels, schools, etc.

7.4.4 Building shape should be simple, building functions should be combined with decorative elements, and the use of decorative elements should be reduced.

7.4.5 Construction project and decoration project should be designed and constructed in integration. Architectural design shall be coordinated with the decoration design, should be synchronized with the decoration design, and shall consider needs of decoration project.

7.4.6 For the indoor space requiring for functional changes in office buildings, shopping malls, etc., the reusable partition walls which are convenient for dismantling, restructuring and reutilization should be used, such as light steel joist plaster board walls, glass walls, boards, etc. The gross area of the room enclosed by un-reusable partition walls shall be in accordance with the requirements of index D6 in Table 4.3.2.

7.4.7 Industrialized building components should be used, such as industrialized awnings, railings, flues, stairs, doors and windows, blinds, integrated kitchens and bathrooms, unit curtain walls, fabricated partition walls, composite exterior walls, integrated ceilings, etc. The industrialized decoration manner should be taken.

7.4.8 The external decoration materials and building constructions with good durability shall be selected, and facilities convenient for maintaining the external elevation of buildings shall be provided. Stainless steel or hot galvanized products with better corrosion resistance should be used for outdoor steel components; window cleaning equipment should be provided for the high-rise buildings with large area of glass curtain wall, and rings for fixing safety belts should be provided for other buildings.

7.4.9 Products with longer life cycle and easy replacement shall be selected for the frequently used movable accessories of buildings, such as hardware fittings, pipe valves, switches, faucets, etc. Equal life cycle shall be considered for combined components. The composition of components with different life cycles should be convenient to dismantle, replace and renew separately.

7.4.10 Except floor heating, the average thickness of the flooring cast-in-place surface of the major use space shall not exceed 50mm, and should not exceed 30mm.

7.4.11 Recycled materials such as plaster boards, metals, glass, wood, etc. should be selected for the external windows and the indoor finishing materials of buildings.

7.4.12 Waste-reuse materials should be selected in priority, such as desulfurized plaster boards, coal ash products, etc.

7.4.13 The existing buildings and structures in the site should be reasonably used; used building materials after construction works and demolition should be used.

7.4.14 Should choose high strength composite materials made of fast-growing timber or bamboo while use timber.

7.4.15 If conditioned, functional building materials should be selected, such as energy storage materials, self-cleaning materials, formaldehyde-free and antibacterial materials, etc.

7.5 Building Acoustical Environment

7.5.1 The allowable indoor noise level of buildings, the airborne sound insulation factor of envelopes and the crash sound insulation factor of floor slabs shall meet the requirements of the current national standard GB 50118 *Code for Sound Insulation Design of Civil Buildings*.

7.5.2 Various rooms should be partitioned according to the different requirements of acoustical environment; noise source space producing higher noise, such as mechanical room should be centralized and kept away from the rooms for working and rest requiring quietness; if arranged next to each other as restricted by conditions, effective techniques for sound insulation and vibration damping shall be taken. The positions of noise sources shall also meet the following requirements:

    1 The noise sources should be arranged underground;

    2 The noise sources such as water pump house, shall not be arranged just under or above residential buildings;

    3 The elevator mechanical room and shaft shall avoied to be arranged next to the rooms requiring quietness;

    4 Accessory rooms producing noise such as washroom, etc. shall be centralized and align in upper and lower floors.

7.5.3 If the noise source space producing higher noise, such as mechanical room, pipe shaft, etc. was arranged next to the space requiring quietness as restricted by conditions, the following techniques for sound insulation and vibration damping shall be taken

    1 The door of the noise source space shall not directly open to the use space requiring quietness;

    2 The walls and floor slabs between the noise source space and the space requiring quietness shall be treated with sound insulation, soundproof doors and windows shall be used;

    3 The wall surface and ceiling of the noise source space should be treated with sound absorption;

    4 Take damping measures for the equipment such as elevators.

7.5.4 For the design of acoustical environment for rooms having particular acoustical requirements, the optimized spatial form and reasonable arrangement of acoustical reflectors and acoustical materials shall be used in priority.

7.5.5 The design of sound absorption is required for the indoor space of places with dense population, such as the corridor and hallway of public building, station, stadium, commercial center, etc. The sound-absorbing ceilings of mineral wool board, perforated board, wood-wool board, etc. should be used, and the techniques of sound-absorbing wall, spatial sound absorber, etc. may be selected.

7.5.6 For buildings adjacent to the urban traffic artery, the performance of sound insulation shall be reinforced for the exterior walls, external windows and outer doors, and for residential buildings, the performance of sound insulation shall be no less than 30dB for the external windows; corridors and the accessory rooms as washroom, etc. should be arranged on the side adjoining to the artery; the facilities as sound barrier, etc. may be used for separating the traffic noise.

7.5.7 For the rooms having requirements for acoustical environment and the rooms having requirements for the crash sound level of floor slabs, the techniques of floating floor slab, elastic surface layer, acoustic ceiling, damping board, etc. may be used for reinforcing the crash sound insulation of the floor slab; when floor heating is used, the insulating layer of floor heating may be combined to reinforce the crash sound insulation of the floor slab; the vibration damping pad of the floating floor slab shall be arranged along the wall and shall not be lower than 40mm.

7.5.8 If lightweight roof is used for the building, the techniques as laying damping materials, providing ceiling, etc. should be used on the roof surface for preventing the noise of rain.

7.6 Daylighting

7.6.1 The planning and architectural design shall meet the requirement of sunshine in national and Beijing code and standard, and the sunlight simulation software shall be used for sunlight analysis. When a residential building has 4 or more living spaces, at least 2 of them shall meet the requirements of the sunlight standard.

7.6.2 Make full use of natural lighting, and shall meet the requirements of the current national standards GB 50352 *Code for Design of Civil Buildings* and GB 50033 *Standard for Daylighting Design of Buildings*. The boundary conditions for light simulation shall meet the requirements specified in Appendix C. 0. 3 and the following requirements as well:

    1 The underground space should have natural lighting;

    2 Should take reasonable shading techniques to avoid glare while use natural lighting.

    3 The sunshade facilities shall meet the requirements of sunlight and daylight standards.

7.6.3 If conditioned, the following techniques should be taken to improve the natural lighting effect in rooms of buildings with insufficient daylight and underground space:

    1 Use Light wells, light scuttles, sunken plazas, semi-basements, etc. ;

    2 Set Light pipes, reflectors, retroreflectors, light collecting devices, prism windows, light guide fiber, etc.

7.6.4 The design of building elevation shall not produce any light pollution to the ambient environment, and the materials as mirror glass or polished metal plate, etc. shall not be used; glass curtain walls shall be made of curtain wall glass with the reflectivity of no larger than 0.30; if glass curtain walls are used on both sides of the urban artery, overpass and viaduct bridge, low-reflectivity glass with the reflectivity of no larger than 0.16 shall be selected.

7.7 Natural Ventilation Environment

7.7.1 The air distribution of natural ventilation shall be designed for buildings, and the space layout, section design and the arrangement of doors and windows shall be in favor of indoor natural ventilation. The indoor ventilation environment should be simulated by computer to optimize the design of natural ventilation. The simulation of natural ventilation shall meet the requirements in Appendix C. 0. 4.

7.7.2 The room plan should adopt the layout favorable for forming draught but avoid the layout of unilateral ventilation.

7.7.3 The positions, directions and opening modes of external windows shall be reasonably designed. For residential buildings, the openable area of the external windows of major use space shall not be less than 1/15 of the room area; for public buildings, the actual openable area of the external windows or transparent curtain walls shall not be less than 5% of the gross area of the exterior wall or curtain wall in the same orientation.

7.7.4 For residential buildings, small-sash windows, natural ventilators, etc. should be selected for convenience of ventilation in heating seasons. When natural ventilators are used, convenient and flexible switch adjusting devices are required for convenience of operation and maintenance, and the natural ventilators should have the functions of filtration and sound insulation.

7.7.5 The following techniques should be taken to increase the internal natural ventilation of buildings:

    1 Techniques for inducing draft should be employed, such as air guiding wall, air catching window, air drafting shaft, ventilating duct, natural ventilator, solar air drafting duct, unpowered hood, etc.; the facilities of air drafting shaft, ventilating duct, etc. should be able to control and close;

2 For building provided with a atrium, openable windows should be arranged at the upper part, and hot-pressing ventilation should be induced by using the chimney effect in suitable seasons; the openable windows shall be able to be closed in winter.

7.7.6 If conditioned, the following techniques should be taken to increase the natural ventilation of the underground space:

1 Semi-basement with direct ventilation may be designed;

2 The basement should be part provided with a sunken courtyard; the sunken courtyard should avoid the influence of automobile exhaust to the upper building;

3 The basement should be provided with vent shafts and window wells.

7.8 Indoor Air Quality

7.8.1 The limits of formaldehyde, benzene, ammonia, radon and other harmful substances in building materials shall meet the requirements of the current national standards GB 18580 *Indoor Decorating and Refurbishing Materials-Limit of Formaldehyde Emission of Wood-based Panels and Finishing Products*, GB 18588 *Limit of Ammonia Emitted from the Concrete Admixtures*, GB 6566 *Limits of Radionuclides in Building Materials* and GB 50325 *Code for Indoor Environmental Pollution Control of Civil Building Engineering*.

7.8.2 The rooms producing peculiar smell or pollutants, such as smoking room, copy room, print room, refuse house, cleaning room, etc., shall be separated from other rooms. Under the premise of no smoking indoors for public buildings, outdoor smoking area shall be arranged and at least 8m from the main entrance of the building.

7.8.3 The main entrance of the public building should be provided with facilities with the function of stopping dust, such as mud scraping pad, mud scraping boarder, etc.

7.9 Other Requirements

7.9.1 The barrier-free design of buildings shall meet the requirements of GB 50763 *Codes for Accessibility Design*.

7.9.2 Consummated facilities and identifiers for barrier-free shall be arranged at the entrances, washrooms, elevators, parking lots, etc., in order to ensure that wheelchair users and those with visual disability can pass through and use smoothly.

7.9.3 Barrier-free residence of newly built residential areas shall not be less than 2%, and barrier-free guest rooms of newly built hotels shall not be less than 1%.

7.9.4 Buildings with centralized dining area shall be provided with an organic refuse collecting area, and the collected refuse shall be sent to the area for centralized processing.

7.9.5 The position for the outer unit of air cooling conditioner shall have good ventilating conditions, short pass of airflow between exhaust air and suction air shall not be allowed, thermal pollution shall be avoided, and clearing and maintaining shall be convenient.

8 Structure Design

8.1 General Requirements

8.1.1 The design working life of structure shall not be less than the requirement in the current national standard GB 50068 *Unified Standard for Reliability Design of Building Structures* and 25 years as well; the resistance and durability of structural components shall meet the corresponding requirements for design working life.

8.1.2 The safty class of building structures shall not be lower than the requirement in the current national standard GB 50068 *Unified Standard for Reliability Design of Building Structures* and should not be lower than Grade II as well.

8.1.3 The structural design shall use the structural system with less resource consumption and small er environmental impact, and shall give full consideration to the techniques for material saving, construction safety, environmental protection, etc.

8.1.4 Structural materials shall be selected in compliance with the principles below:

1 Materials with less resource consumption and small er environmental impact should be selected; recyclable and reusable materials are preferred, and the service efficiency of the materials should be improved;

2 Materials with high performance and high strength should be used; premixed concrete shall be used as the cast-in-situ concrete;

3 Materials produced within 500km from the construction site should be selected;

4 Never use any material with high energy consumption or pollutants over-discharged;

5 Never use any material restricted or eliminated by the state or Beijing.

8.1.5 The optimization design shall be provided for the structural design:

1 Optimization design of seismic performance objectives of structure;

2 Optimization design of structural system;

3 Comparison and selection optimization design of structural materials (material types and strength grade);

4 Optimization design of structural component layout and section.

8.2 Design of Main Structure

8.2.1 The design load value of structure may be properly increased for newly built buildings.

8.2.2 The structural arrangement should improve the adaptability to the building layout.

8.2.3 The structural scheme shall meet the requirements for seismic concept design, and the seriously irregular structural scheme shall not be used; for particularly irregular structures, the seismic performance objectives shall be determined by consultation with the owner and relevant experts.

8.2.4 Under the conditions of ensuring safety and durability, the optimization design of the structural system shall be in accordance with the following requirements:

1 Any structure difficult to implement or Over-limit seismic structure formed by irregular

building shape should not be used;

 2 The structural system with less material consumption shall be selected according to the loading characteristics;

 3 Seismically isolated or energy-dissipated structure should be preferred for Class-A buildings; if conditioned, seismically isolated or energy-dissipated structure shall be used for Class-B buildings;

 4 In high-rise and long-span structures, the steel structure and the steel reinforced concrete composite structure shall be reasonably used.

8.2.5 Materials shall be selected in accordance with the following requirements:

 1 Premixed mortar shall be used;

 2 High strength steel bars and high strength steel shall be reasonably used: the proportion of amount of high strength steel bars for high-rise reinforced concrete structures and the proportion of amount of high strength steel for high-rise steel structures shall meet the requirements in 4.3.2;

 3 High strength concrete shall be reasonably used: the concrete strength grade of the vertical components at the lower part of high-rise building structures above 60m shall meet the requirements in 4.3.2;

 4 Structural materials of high industrialized level should be used.

8.2.6 The optimization design of structural components shall be in accordance with the following requirements:

 1 The section optimization design shall be provided for the vertical components of high-rise structures and the horizontal components of long-span structures;

 2 For long-span concrete structures, the prestressed floor system, cast-in-situ concrete hollow floor system, sandwiched floor system, etc. should be reasonably used;

 3 For steel structure components controlled by strength, high strength steel should be used; for steel structure components controlled by stiffness, the component layout should be adjusted;

 4 Components with obvious effect of material saving and high industrialized level should be reasonably used.

8.3 Design of Building Foundation

8.3.1 The design of building foundation shall stick to the principles of using local materials, protecting environment, saving resources and improving efficiency in combination with the actual conditions in Beijing, and shall give comprehensive considerations to the factors of construction conditions, site environment, project costs, etc. according to the investigation report, structural characteristics and architecture function.

8.3.2 According to the conditions of upper structures, foundation should adopt natural foundation, ground treatment and pile foundation in order.

8.3.3 The ground treatment should adopt the methods of cushion, cement-fly ash-gravel piles (CFG pile) and compaction piles. If environment permits, dynamic compaction and rammed soil-cement piles may also be used.

8.3.4 The collaborative analysis and design of building foundation shall meet the following requirements:

 1 For high-rise buildings, the combined action of foundation with the upper structure should be considered for collaborative design;

 2 For the pile foundation of settlement control, the synergism of pile cap pile and soil should be considered;

 3 The optimization design of raft foundation should be conducted according to the results of collaborative computing.

8.3.5 Cast-in-situ bored piles should take the post grouting technique to improve the side resistance and end resistance.

8.3.6 For urban artificial fill, the processed industrial residues, inorganic construction wastes and plain fill shall be selected nearby as the foundation of multistoried buildings, and it shall meet the requirements of relevant codes.

8.4 Design of Renovation and Extension Structure

8.4.1 **For renovation and extension projects, the original building structures shall be utilized as much as possible based on the appraiser of structural reliability. As required by the appraiser of structural reliability, they shall be used continuously according to the assessed working life after employing the necessary techniques of strenthening and maintenance.**

8.4.2 For renovation and extension projects, the structural components of the original building should be reserved and shall be treated with necessary maintenance and strengthening.

8.4.3 If the stiffness of the building and the bearing capacity of the structural components can't meet the requirements of the current codes for structural design due to the building functional changes, story-adding, renovation, extension, etc. , or the class of seismic protection of building, optimized structural system and the strenthening scheme for structural components shall be adopted, and the optimized strenthening scheme for structural system should be preferred.

8.4.4 The material-saving, energy-saving and environmental-protecting reinforcement techniques shall be used for the reinforcement of the structural system or components.

8.4.5 The structural renovation and extension shall make full use of the reusable structural materials produced by building construction, building demolition and site clearing.

9 Design of Water Supply and Sewerage

9.1 General Requirements

9.1.1 The scheme for water resource planning of construction projects shall be drafted at the conceptual design stage, so as to make overall planning and comprehensive utilization of various water resources. The water resource planning shall include the contents of the comprehensive utilization of nontraditional water sources, such as reclaimed water, rainwater, etc.

9.1.2 The industrial waste heat, exhaust heat and the heat of condensation shall be selected in priority as the heat source of central hot water supply; if conditioned, geotherm and solar energy may be utilized to prepare hot water. Besides, it shall meet the relevant requirements of DB11/891 *Design Standard for Energy Efficiency of Residential Buildings* and DB 11/687 *Design Standard for Energy Efficiency of Public Buildings* of Beijing.

9.1.3 The arrangement of facilities and pipelines for water supply and sewerage shall not produce any noise pollution to indoor and outdoor environment.

9.1.4 The following building drainage shall be separately drained to the building for water treatment or water reclamation structure:

    1 Washes containing a large amount of grease from staff dining hall and commercial restaurants;

    2 Wash water of automatic mechanical car wash;

    3 Hospital wastewater containing plenty of pathogenic bacteria and over-discharged radioactive elements;

    4 Drainage exceeding 40℃ from boiler, water heater and other heating equipment;

    5 Domestic drainage used as the water source of reuse water;

    6 Poisonous and harmful waste water from laboratory.

9.2 Water Supply System

9.2.1 For the water saving design of water supply system, techniques shall be employed to make comprehensive use of various water resources adjusting measures to local conditions, such as municipal water supply, municipal recycled water, rainwater, building reclaimed water, etc. When the nontraditional water sources are used, the supplied water quality index shall be reasonably determined according to the use functions.

9.2.2 In the design of water saving planning, the average daily rated water consumption shall take the lower value of the water index in GB 50555 *Standard for Water Saving Design in Civil Building*.

9.2.3 When municipal water supply system and municipal recycled water system are adopted, the water pressure of the urban pipe network for municipal water supply shall be fully utilized. If pressurized water supply is required, the energy-saving water supply technology by the pressure superposed water supply shall be adopted in priority. For multistory buildings and high-rise buildings, the vertical division zone shall be reasonably determined for the systems of water supply, reclaimed water and hot water; for public buildings, the water supply pressure of each water taking position shall not be larger than 0.15MPa; for residential buildings, the water supply pressure of each water taking position shall not be larger than 0.20MPa and no less than minimal pressure required by the water devices.

9.2.4 In case of less water consumption and scattered water taking positions for hot water, local hot water supply systems should be used; in case of more water consumption and centralized water taking positions for hot water, central hot water supply systems shall be used, and the consummated hot water circulation system shall be provided. The hot water system shall be arranged in accordance with the following requirements:

1 When the residential building is provided with central hot water supply, the circulation of main pipe and stand pipe shall be provided, and the draining time of the water temperature reaching the designed water temperature shall not be larger than 15s at each water taking position;

2 For public buildings such as hospitals, hotels, etc., the draining time of the water temperature reaching the designed water temperature shall not be larger than 10s at each water taking position;

3 For bathhouses, the water saving techniques of fixed quantity or fixed time, etc. shall be employed for the showering hot water system.

9.2.5 The central hot water supply system shall have the techniques for ensuring the balance of water supply pressure between cold water and hot water in each water taking position; the difference in water supply pressure between cold water and hot water shall not be larger than 0.02MPa in the most unfavorable water taking position, and shall be in accordance with the following requirements:

1 The cold water and hot water supply systems shall be in unified zoning;

2 When it is difficult to keep the unified zoning of cold and hot water systems, the technique of providing adjustable pressure reducing valves for water distribution branches should be employed to reduce pressure, so as to ensure the pressure balance between cold water and hot water for the system.

9.2.6 When the following systems are adopted, the water saving techniques of water circulation or water recovery shall be employed, and the systems shall be in accordance with the following requirements as well:

1 Cooling water must be recycled;

2 For swimming pools, aquatic entertainment pools (excluding that for children), etc., the recirculated water supply system shall be used, and the discharged wastewater should be recovered for reuse;

3 Steam condensate shall be recovered for reuse or recycling, but never be drained directly;

4 Nontraditional water resources should be used as water for automobile washing; if tap water is used, the wash water shall be recycled;

5 For large-scale buildings equipped with central air-conditioning system, separative devices for recovery and reuse of air-conditioning condensate water should be provided.

9.2.7 Never use the municipal water supply or self-owned underground well water supply as the water-scape. It shall be in accordance with the following requirements:

1 The nontraditional water sources of rainwater, recycled water, etc. shall be used as the water-scape;

2 The supplemented water amount for waterscape shall be in balance with the amount of recovered rainwater and the amount of building reclaimed water;

3 The waterscape shall be used after circular processing.

9.2.8 For civil buildings, water meters shall be arranged in the following positions of the pipelines of water supply, hot water, reclaimed water, pipe portable water, etc.:

1 On the water supply pipes of each residential unit and those used for different purposes of landscape, irrigation, etc. for residential buildings;

2 On the water supply pipes used for different purposes and those in different units of payment for public buildings.

9.3 Water Saving Equipment and Sanitary Ware

9.3.1 Water nozzles, showers, domestic washing machines, water closets, flush valves, etc. shall meet the requirements of the current industrial standard CJ 164 *Domestic Water Saving Devices*.

9.3.2 The efficient and water-saving irrigation modes of sprinkling irrigation, micro-irrigation, etc. shall be used for greening watering.

9.3.3 Finished products of cooling tower shall select the products with higher cooling efficiency but less water loss. The ratio of water loss of cooling towers shall be less than 0.01%

9.4 Utilization of Non-traditional Water Source

9.4.1 Reclaimed water facilities must be designed and constructed for the following newly built buildings:

1 Hotels, restaurants, apartments, etc. with the building area of more than 20,000m^2;

2 Organs, research units, universities, large-scale buildings for culture and physical training, etc. with the building area of more than 30,000m^2;

3 Residential areas and centralized construction area with the building area of more than 50,000m^2 or the recoverable water amount of larger than 150m^3/d.

9.4.2 The recycled water of urban or regional centralized recycled water plants shall be used in priority as the reclaimed water resources of residential districts.

9.4.3 The analysis of water quantity balance and technical economy shall be performed according to the water quality and water quantity of the available raw water and the purposes of the reclaimed water, so as to reasonably determine the water resources, system form, processing technology and scale of reclaimed water.

9.4.4 The paths for rainfall runoff on ground surface and roof shall be reasonably planned so as to reduce the surface runoff but increase the infiltration capacity of rainwater. The schemes for rainfall accumulation and utilization shall be reasonably determined by economic and technical comparisons. Refer to Appendix B.0.3 Rainfall Conditions for Design.

9.4.5 In the process of using nontraditional water source, techniques for ensuring the safety in use must be employed, and the usage shall be in accordance with the following requirements:

1 Pipelines of nontraditional water sources must not be connected with the water supply pipelines of drinking water;

2 Obvious identifiers of nontraditional water source shall be arranged in the positions of cisterns (water tanks), valves, water meters, turncocks and water outlets;

3 Locking devices shall be arranged on the turncocks of nontraditional water source in public places and the water outlets for greening.

10 Heating, Ventilation and Air-conditioning Design

10.1 General Requirements

10.1.1 For the design of heating systems and air-conditioning systems, the heating load as well as termwise and hourly cooling load must be calculated for each air-conditioned room or each air-conditioned zone. When ground-source heat pump and other renewable energy sources, and the new energy and new forms of energy-saving system such as combined cooling, heating and power (CCHP) supply system and energy-storage system are used, the annual dynamic load and energy consumption should be simulated, and the energy consumption and technical economy should be analyzed to select the reasonable heat and cold sources and the forms of heating and air-conditioning systems. It shall also meet the following requirements:

1 For public buildings, the basic data of thermal parameters of envelope, number of users, lighting power density, indoor equipment, work and rest modes, etc. used for calculating the cooling load and heating load of heating, ventilation and air-conditioning shall be kept in coordination with those in other majors;

2 The Outdoor calculating parameters for heating, ventilation and air-conditioning's design shall be determined in accordance with Appendix B.0.1; the indoor calculating parameters for heating, ventilation and air-conditioning design should not be higher than the standard specified in 4.1.2 of DB 11/687-2009 *Design Standard for Energy Saving of Public Buildings* of Beijing;

3 Under the design working conditions, the fresh air volume shall not exceed the minimum fresh air volume as specified in GB 50736 *Design Code for Heating Ventilation and Air Conditioning of Civil Buildings*;

4 When the annual energy consumption of buildings is simulated, the simulation setting shall meet the requirements in Appendix C.0.2.

10.1.2 The zoning and forms of heating, ventilation and air-conditioning systems shall be designed according to the room functions, building orientation, building space form, service time, property attribution, requirements for control and adjustment, internal and external areas, characteristics of annual cooling and heating load, etc.

10.1.3 Central cooling air-conditioning system shall not be used for residential buildings.

10.1.4 The rooms which may produce fumes, peculiar smell and other pollutants, such as kitchen, washroom, smoking room, refuse house, copy room, etc., shall be provided with exhaust air systems so as to maintain the negative pressure in the rooms.

10.1.5 Except the buildings having special requirements, such as kindergartens, etc., the heating radiators shall not be installed in a concealed mode for residential buildings and public buildings.

10.1.6 For heating, ventilation and air-conditioning systems, the equipment with low noise and low vibration shall be selected; the corresponding techniques for noise elimination, vibration isolation and vibration damping shall be employed according to the requirements for technology and use function, magnitudes of noise and vibration, frequency characteristics, mode of transmission, the standard allowed for noise vibration, etc.

10.2 Energy Transportation and Distribution

10.2.1 For air-conditioning and ventilating systems, the power consumption of unit air volume shall meet the requirements of 4.3.4 in DB 11/687 - 2009 *Design Standard for Energy Saving of Public Buildings* of Beijing. When the cold and hot water circulating pump is selected for centralized heating and cooling system, the ratio of electricity consumption to transferred cold (heat) quantity [EC(H)R] of the water circulating pump shall be calculated and marked in the design document of working drawings, and it shall meet the requirement in the formula below as well:

$$EC(H)R = \frac{0.003096 \Sigma(G \cdot H/\eta_b)}{Q} \leqslant \frac{A(B+\alpha \Sigma L)}{\Delta t} \quad (10.2.1)$$

Where: $EC(H)R$ ——ECR refers to the electricity consumption to transferred cooling quantity ratio of cold water circulating pump, whereas EHR refers to the electricity consumption to transferred heat quantity ratio of hot water system;

G ——Designed water flow of each operating water pump, m³/h;

H ——Designed lift head of delivery to each operating water pump, mH₂O;

η_b ——Efficiency of the operating water pump corresponding to the designed operating point;

Q ——Designed cold/heat load, kW;

Δt ——Designed temperature difference between supplied water and return water (℃); the value shall be selected according to Table 10.2.1-1;

A ——Calculating coefficients related to the flow of water pump; the value shall be selected according to Table 10.2.1-2;

B ——Calculating coefficients related to the water resistance of plant room and user; the value shall be selected according to Table 10.2.1-3;

ΣL —— For heating system, it refers to the total length of outdoor main line (including water supply and return pipes), m; for air-conditioning system, it refers to the total length of transmission from the plant room of heat and cold sources to the water supply and return pipes of the farthest user of the system, m;

α ——Calculating coefficients related to ΣL; the value shall be selected according to Table 10.2.1-4.

Table 10.2.1-1 Δt Values (℃)

Hot Water System for Heating	Chilled Water System for Air-conditioning	Hot Water System for Air-conditioning
Designed Temperature Difference Between Supplied Water and Return Water	5	15

Table 10.2.1-2 Values of Calculating Coefficients (A) Related to the Flow of Water Pump

Designed Flow of Water Pump (m³/h)	$G \leqslant 60$	$60 < G < 200$	$G \geqslant 200$
A Value	0.004225	0.003858	0.003749

Note: When water pumps of different flows are operating in parallel, the A Value shall be selected according to the maximum flow of a single unit.

Table 10.2.1-3 Values of Calculating Coefficients (B) Related to the Water Resistance of Motor Room and User

System Components		B Value for Four-pipe Single-cold and Single-hot pipeline	B Value for Two-pipe Hot Water Pipeline	Heating Pipeline
Primary Pump	Chilled Water System	28	—	—
	Hot water system	22	21	20.4
Secondary Pump	Chilled water system	33	—	—
	Hot water system	27	25	24.4

Note: For the cold water system of multistage pump, the B value may be added by 5 once one-stage pump is increased; for the hot water system, the B value may be added by 4 once one-stage pump is increased.

Table 10.2.1-4 Values of Calculating Coefficients (α) Related to $\sum L$

System	Range of Pipeline Length $\sum L$(m)		
	$\sum L \leqslant 400$	$400 < \sum L < 1000$	$\sum L \geqslant 1000$
Four-pipe Chilled Water System	$\alpha = 0.02$	$\alpha = 0.016 + 1.6/\sum L$	$\alpha = 0.013 + 4.6/\sum L$
Four-pipe Hot Water System	$\alpha = 0.014$	$\alpha = 0.0125 + 0.6/\sum L$	$\alpha = 0.009 + 4.1/\sum L$
Heating Water System	$\alpha = 0.0115$	$0.003833 + 3.067/\sum L$	0.0069

Note: In the hot water system for heating, $\sum L$ (m) refers to the total length of outdoor main line.

10.2.2 The temperature design for supplied water and return water of heating, ventilation and air-conditioning systems shall meet the following requirements:

1 Except the temperature & humidity independent processed air-conditioning system and free cooling system in winter, the temperature difference between supplied water and return water of the chilled water system for electric cooling shall not be less than 5℃;

2 Except using waste heat or heat pump system, the temperature difference between supplied water and return water for the hot water system for air-conditioning should not be less than 15℃;

3 For the central heating system using a radiator at the terminal, the temperature of the supplied water shall be no higher than 90℃ but should not be lower than 65℃, and the temperature difference between supplied water and return water should not be less than 20℃;

4 When ice-storage cold source for air-conditioning is used or chilled water lower than 4℃ is available, the chilled water system with large temperature difference for air-conditioning should be used;

5 The cooling radius of the chilled water system for air-conditioning should be controlled within 300m. When the cooling radius exceeds 300m and it is reasonable by technical and economic comparisons, the chilled water system with bigger temperature difference and small flow should be adopted.

10.2.3 The hydraulic balance shall be calculated for the air-conditioning water system, and the balancing techniques of adjusting pipe diameter and setting resistance valve may be employed according to the computed results.

10.2.4 For the air-conditioning and ventilating systems, the action radius should not be too large, whereas the position of the air-conditioning unit shall not be far away from the room or space to be serviced, and the techniques for vibration isolation, sound insulation and noise elimination shall be employed. The positions of the fresh air inlet and air exhaust shall avoid the pipeline of the ventilating system from being too long. The vertical load of the air-conditioning and ventilating systems should not exceed 10 floors.

10.2.5 The ducts design of air-conditioning and ventilating systems shall be in accordance with the following requirements:

1 For the rectangular air duct of air-conditioning and ventilating systems, the length-width ratio of the cross section should not be larger than 4 and shall not be larger than 8; or the specific frictional resistance should not be larger 2 Pa/m and shall not be larger than 4 Pa/m;

2 For the ducts of air-conditioning and ventilating system, bends and tee joints with less resistance loss shall be used at the elbows and branches;

3 For the ducts of air-conditioning and ventilating systems, control valves, test holes, access holes and cleaning holes for air duct shall be provided for debugging and maintenance.

10.3 Heat and Cold Source

10.3.1 For civil buildings, the waste heat of power plants or other industrial waste heat and urban heat shall be used in priority as the heat source of heating. When other renewable energy sources may be used for reducing the consumption of conventional energy resources; if they are reasonable after technical and economic comparisons, the renewable energy sources should be selected in priority.

10.3.2 The installed capacity of the heat and cold source shall be determined on the basis of the cold and heating load calculated under the designed conditions, and the recovered cold and heat of energy shall be deducted. The installed capacity and number of units shall be reasonably matched following the principle of situating in the status of relatively higher efficiency when the system is under the frequently part load.

10.3.3 When the building has higher plot ratio, matched cheating load, power load and cold load as well as stable heating load, if the average ratio of multipurpose utilization of energy for the CCHP supply system is higher than 80% as analyzed by calculating the annual heating, power and cold loads, and it has reasonable technical economy, the distributive CCHP supply system operating in the mode of determining power by heating load may be adopted.

10.3.4 The projects adopting regional cooling shall meet the following requirements:

1 Higher plot ratio, small cooling radius, dense cooling load for air-conditioning, satisfying the conditions for ice storage, annual energy consumption calculated and analyzed, and being reasonable in economic and technical comparison;

2 Being reasonable in economic and technical comparison, adopting multistage pump, large temperature difference but small flow, variable flow control, and employing the techniques and measures of reinforcing thermal insulation, etc.

10.3.5 When waste heat steam and fume or waste heat hot water no lower than 80°C are available, the absorption cooling should be adopted.

10.3.6 When the peak and valley loads for air-conditioning have a wide difference and the economic and technical comparison has reasonable results in combination with the policy of electric price differences between peak and valley hours, the cold-storage cold source for air-conditioning may be adopted.

10.3.7 For water chilling (heat pump) units, the coefficient of cooling performance under the rated operating conditions shall be one grade higher than that required in DB 11/687 *Design Standard for Energy Saving of Public Building* in Beijing. For boilers, the thermal efficiency under the rated operating conditions shall meet the requirement for the limit value in the Appendix A of TSG G0002-2010 *Supervision Administration Regulation on Energy Conservation Technology for Boiler*.

10.3.8 For electrical cooling units (including ground-source heat pumps), the summated refrigerating coefficient of performance (SCOP) under the nominal operating conditions shall be superior to the requirements in the current DB 11/687 *Design Standard for Energy Saving of Public Building* of Beijing.

10.3.9 Under the designed operating conditions in winter, the air-source heat pump units should not be used as the heating equipment in winter when the coefficient of performance (COP) is lesser than the values below:

1 Cold and hot air unit for air source heat pump: less than 1.80;

2 Cold and hot water unit for air source heat pump: less than 2.00.

10.3.10 When the ground-source heat pump and other renewable energy sources are used as the heating and cold source for heating, ventilation and air-conditioning, the ratio of contribution for the renewable energy sources shall be calculated according to the requirements in 4.2.2.

10.3.11 For the public buildings having larger internal area and a plenty of stable waste heat all year round, if reasonable in the economic and technical comparison, the air-conditioning systems capable of recovering waste heat shall be used, such as water loop heat pump, etc.

10.3.12 In winter, the indoor waste heat shall be eliminated by using outdoor fresh air in priority, or the cooling tower shall be used for providing cold water for the internal area of buildings.

10.4 Control and Detection

10.4.1 For large-scale public buildings, the automatic control system for heating, ventilation and air-conditioning shall include complete or partial contents of detection and control, etc., such as parameter detection, indication for status and fault of parameters and equipment, equipment linkage control and automatic protection, automatic switching of operating conditions, energy metering, automatic regulation and control, central monitoring and management, etc.

10.4.2 For the design of automatic control system for heating, ventilation and air-conditioning, the strategies for the operation of partial load and the operation of each functional zone shall be defined. The design of heating, ventilation and air-conditioning equipment and system shall meet the requirements for layering, zoning and time-sharing control, so as to realize the functions of controlling and regulating the indoor air-conditioning in different rooms.

10.4.3 As required by the management for property attribution and operation, energy metering devices shall be installed for heating, ventilation and air-conditioning system to metering the cold and heat quantities. For large-scale public buildings, the metering of power consumption by items for the heating, ventilation and air-conditioning equipment and system shall meet the following requirements:

  1 The power consumption of equipment in the plantroom of heat and cold sources shall be metered by items according to heat source equipment, hot water circulating pump, cold source equipment, fan of cooling tower, cooling water pump, chilled water pump, etc.;

  2 The power consumption of terminal air-conditioning equipment should be metered by items according to air-handling unit/fresh air unit/fun, fan-coil unit, split air-conditioner, etc.;

  3 The power consumption of equipment for heat and cold source of energy-storage system shall have the function of metering at different time periods.

10.4.4 For the functional space with dense population, the equipment shall be capable of operation to change the fresh air volume according to the changes of population or the CO_2 concentration in winter and summer; full air-conditioning system shall change the overall air volume output according to the changes in partial loads.

10.4.5 For underground garages, the ventilating system shall be capable of adjusting and controlling the operation of the ventilating system according to the changes in service time, different times for the frequent passing through of automobiles, pollutant concentration in automobile exhaust, etc.

10.4.6 The water systems of air handling unit and fan-coil unit shall be designed according to the fixed water temperature difference and the operating mode of variable water flow.

11 Building Electrical Design

11.1 General Requirements

11.1.1 This chapter is applicable to the building electrical design of power supply and distribution system of 10(6)kV and below for civil buildings.

11.1.2 At the scheme design stage, reasonable schemes for power supply and distribution system and intelligent system, the energy-saving and efficient electrical equipment meeting the functional requirements shall be selected in priority, the energy saving technology shall be reasonably applied, and energy saving and high efficiency shall be taken as the major technical and economic indicators for comparing scheme design.

11.1.3 At the scheme design stage, the renewable energy sources in the site shall be evaluated. In case of reasonable technology and economy, solar power generation, wind power generation, CCHP supply, etc. should be selected as the supplementary power energy, and the combined-grid power generation system should be used.

11.1.4 The type selection and installation of electrical equipment, such as transformer, diesel-driven generator, wind-driven generator, etc., shall avoid producing noise pollution to the building and surrounding. The arrangement of the fume exhaust in the room of diesel-driven generator shall meet the requirement of Item 4 in 6.1.3 of JGJ 16 - 2008 *Code for Electrical Design of Civil Buildings*.

11.2 Power Supply and Distribution System

11.2.1 The design of power supply and distribution system shall improve the operating efficiency of the entire power supply and distribution system on the basis of satisfying safety, reliability, technical rationality and economy.

11.2.2 The power substation should be close to the load center. If conditioned, the range of power supply for the power substation of large-scale public buildings should not exceed 200m.

11.2.3 For the design of power supply and distribution system, power load and reactive power shall be calculated, and the capacity and quantities of transformer shall be reasonably selected.

11.2.4 When the harmonic waves of the power supply and distribution system or equipment exceed the requirement for the limit value of harmonic wave in the current national or local standards, the nature of the harmonic wave source, the harmonic wave parameters, etc. shall be analyzed, and the corresponding techniques for suppression and treatment of harmonic waves shall be employed. Filtering devices should be provided for the equipment or places with great harmonic wave interference in buildings.

11.2.5 The cross section of the power cables of 10kV and below shall be designed in consideration of the technical conditions, operating conditions and economic current. The cross section of economic current shall be selected in accordance with the relevant requirements in Appendix B of GB 50217 - 2007 *Code for Design of Cables of Electric Engineering*.

11.3 Lighting

11.3.1 The reasonable standard values of illumination shall be determined according to the factors of project scale, functional characteristics, construction standard, requirement for visual operation, etc.

11.3.2 The natural lighting shall be reasonably used according to the requirements of each place for lighting in buildings, and shall meet the following requirements:

1 For the areas having the conditions of natural lighting or facilities for natural lighting, the lighting design shall be combined with the conditions of natural lighting for the arrangement of artificial lighting;

2 The areas having natural lighting shall be independently controlled by zoning, and should be provided with the devices capable of automatic control or regulation along with the changes in the outdoor natural lighting;

3 For the public area of residential buildings having natural lighting, the lighting should be provided with the control device integrating one or more of voice control, light control, timing control, inductive control, etc.

11.3.3 For the rooms and places having the standard value of illumination of 300lx or above and applicable for local lighting, the lighting mode of combining common lighting with local lighting should be selected.

11.3.4 According to the factors of functional characteristics of buildings, construction standard, requirements for management, the lighting shall be controlled by means of combining scattered control with centralized control and combining manual control with automatic control, and shall meet the following requirements:

1 For large space, such as landscape, parking garage, open type office, hall, etc., centralized control should be selected for common lighting, whereas scattered control should be selected for local lighting;

2 For the areas where people do not stay for long, such as meeting room, washroom, etc., inductive control devices may be installed;

3 For elevator hall, corridor, stair cases, etc., timing or inductive control devices should be installed;

4 For the public buildings with higher requirement for lighting environment or complicated function and large-scale public buildings, intelligent lighting control system should be installed independently, and it shall have the functions of light control, time control, induction, communication with the Building Management System, etc.

11.3.5 For the areas where people work or stay for long, the light source for lighting shall have the color rendering index (Ra) of no less than 80.

11.3.6 Except the places having special requirements, the lighting design shall select efficient light source for lighting, efficient lamps and energy-saving accessory devices.

11.3.7 For various rooms or places, the lighting power density should not be higher than the target value specified in the current national standard GB 50034 *Standard for Lighting Design of Buildings*.

11.4 Electrical Equipment

11.4.1 For distribution transformers, transformers with D, yn11 connection mode shall be selected, and the energy-saving products with low loss and low noise shall be selected. The no-load loss and load loss of distribution transformers shall not be higher than the evaluation value of energy saving specified in the current national standards GB 20052 *The Minimum Allowable Values of Energy Efficiency and the Evaluating Values of Energy Conservation for Three-phase Distribution Transformers*.

11.4.2 Low-voltage AC motors shall select high performance motors, and the energy efficiency shall meet the evaluation value of energy saving specified in the current national standard GB 18613 *Minimum Allowable Values of Energy Efficiency and The Energy Efficiency Grades for Small and Medium Three-phase Asynchronous Motors*.

11.4.3 Elevators equipped with efficient motor and advanced control technology shall be used. Escalators and moving pavements shall be provided with energy-saving dragging devices and energy-saving control devices, and should be equipped with the inductive sensors capable of automatically controlling the escalators and moving pavements.

11.4.4 When two or more elevators are arranged in a centralized mode, the control system shall have the functions of centralized regulation and group control according to programming.

11.5 Metering and Intelligentization

11.5.1 For residential buildings, the power consumption shall be metered by households and purposes; in addition to the relevant professional requirements, the metering shall also be in accordance with the following requirements:

    1 A power metering device shall be provided for each household;

    2 A power metering device shall be provided for the lighting of public area;

    3 Independent power metering devices by items shall be respectively provided for elevators, heating stations, reclaimed water equipment, water supply equipment, sewerage, air-conditioning equipment, etc.;

    4 Independent metering devices by items shall be provided for the power generation of renewable energy sources.

11.5.2 For public buildings, the metering of power consumption shall be set according to purposes, property attribution, operating management and the requirements of relevant major; for the office buildings of state organs and large-scale public buildings, the metering by items shall also meet the relevant requirements in DB 11/T 624 *Technical Requirements for Measurement of Electricity Usage by Category in Office Buildings of Public Agencies*, and shall be in accordance with the following requirements:

    1 A main electricity meter shall be arranged at the entrance of each independent building;

    2 Independent power metering devices by items shall be provided for lighting, elevators, refrigeration plate, heating stations, air-conditioning equipment, reclaimed water equipment, water supply equipment, sewerage, lighting for landscape, kitchen, etc.;

    3 A power metering device shall be provided for each rent unit for office or business;

    4 The power consumption of office equipment, lighting shall be metered by items or households for office buildings;

  5 When mechanical ventilation is adopted for the non-air-conditioning area of basement, an independent power metering device should be installed;

  6 Independent metering devices by items shall be provided for the power generation of renewable energy sources;

  7 For large-scale public buildings, independent power metering devices by items shall be provided for the ventilating and air-conditioning equipment in special places such as kitchens, computer rooms, etc. ;

  8 For large-scale public buildings, the metering by items of power consumption of heating, ventilation, air-conditioning and cooling system shall meet the requirements of 10. 4. 3.

11.5.3 The metering devices should be arranged in a relatively centralized place. When conditionally restricted, remote meter reading system or card meter device may be selected.

11.5.4 The power metering devices shall be selected in accordance with the following requirements:

  1 The intelligent meters monitored by the distribution monitoring and control system should include the parameters of voltage, current, electrical quantity, active power, reactive power, power factor, etc. ;

  2 For the electrical meters in key positions, advanced fully electronic watt-hour meters should be selected;

  3 Both prepayment IC meter and remote meter shall be approved by the Department of Measurement & Testing.

11.5.5 The data acquisition standard for energy consumption of buildings shall be in accordance with the relevant requirements in JGJ/T 154 *Standard for Energy Consumption Survey of Civil Buildings*.

11.5.6 For buildings equipped with intelligent system, the subsystem configuration shall be matched in accordance with Appendixes A ~ J of GB/T 50314 - 2006 *Standard for Design of Intelligent Building*; for residential buildings, the design of intelligent system shall also meet the requirements for basic configurations in CJT 174 *Technological Classification of Community Intellectualization System*.

11.5.7 Large-scale public buildings shall have the functions of monitoring and controlling equipment for public lighting, air-conditioning, water supply and sewerage, elevators, etc. , and should be provided with building intelligent system integration.

11.5.8 The office buildings of state organs and large-scale public buildings shall be provided with building equipment energy management system, and shall have the functions of real-time statistics, analysis and management for energy. Other public buildings shall have the functions of monitoring and statistical management for the energy consumption of major energy-intensive equipment.

11.5.9 The indoor air quality monitoring system should be arranged in accordance with the following requirements:

  1 The public space with people gathering or major functional rooms with high density of people should be equipped with CO_2 concentration detector and display device to realize real-time alarm when the CO_2 concentration exceeds the standard;

  2 Underground parking garages should be equipped with CO concentration detector and display device to realize real-time alarm when the concentrations of CO and pollutants exceed the standard;

3 When the above places are equipped with mechanical ventilation system or central air-conditioning system, the ventilating and air-conditioning equipment of the relevant areas should be linked and controlled according to the results detected by the detector.

12 Landscape Design

12.1 General Requirements

12.1.1 The landscape design shall comply with the design principle of the integral sustainable development of economy, environment and society, shall meet the planning and design requirements, and shall be coordinated with the building complex and roads in the site.

12.1.2 The landscape design shall comply with the design principle of adjusting measures to local conditions, and shall make full use of the existing landform, water system and vegetation on site for united design, so as to realize the design purpose of energy, land, water and material saving and environmental protection for green buildings.

12.1.3 The master plan of the landscape design shall give comprehensive consideration to optimizing the wind environment, acoustical environment, light environment, thermal environment, air quality, visual environment, olfactory environment, etc. And various factors of landscape shall be designed in interrelation.

12.2 Greening

12.2.1 The existing trees on site shall be fully protected and utilized.

12.2.2 For planting design, native plants adapting to the regional climate and soil conditions shall be selected, and the native plant index should not be less than 0.7. It is suitable to select the harmless plants with strong weather resistance, easy maintenance as well as fewer pest and disease damages. Refer to Appendix B.0.5 for the plant types.

12.2.3 For planting design, the plants shall be matched according to their ecological habits, and it should meet the following requirements:

    1 Multiple varieties of plants shall be planted by reasonable matching. If the residential area has the land area of no more than 50,000m^2, no less than 30 varieties of native plants shall be planted; if the residential area has the land area of 50,000m^2 ~ 100,000m^2, no less than 35 varieties of native plants shall be planted; if the residential area has the land area of no less than 100,000m^2, no less than 40 varieties of native plants shall be planted;

    2 Multi-layer planting dominated by phytocoenosium shall be adopted, and it shall combine trees and shrubs with grass and ground covers;

    3 More trees and shrubs should be planted in green space, but the planting areas of turf and lawn without tree cover shall be reduced, and no less than three trees shall be planted in every 100m^2 of green space.

12.2.4 For green roof design, the plants with slow growth and strong resistance shall be selected rather than those whose root system has strong penetration. For intensive green roof, also known as rooftop garden design, small trees and shrubs should be reasonably matched to form multi-layer greening. Refer to Appendix B.0.5 for the plant varieties for green roof.

12.2.5 Vertical greening should be dominated by vines in ground culture and pot culture. Different kinds of vines may be selected according to different environment to be attached, and vertical greening should be provided for exterior walls of buildings, site enclosure, rails, roof, entrance and exit of garage, ventilating facilities of subway, guardrails along roads, street furniture, etc.

12.2.6 For the site of greening in soil, sunken green space should be arranged according to the local conditions. Water-resistance plants shall be selected for the planting design in the sunken green space. The ratio of the sunken green space in total site area should not be lower than 50%.

12.2.7 It is preferred to improve, the acoustical environment on site through planting design, and tall trees and shrubs should be planted around the noise source to form the sound-proof barrier.

12.2.8 The planting design should be helpful for improving the light environment on site, and should meet the following requirements:

　　1 Tall trees shall be planted to reduce the glare pollution caused by the reflected light from the building facades;

　　2 Deciduous broad-leaf trees shall be planted aroundthe open space.

12.2.9 The planting design should be in favor of optimizing the thermal environment on site, and should meet the following requirements:

　　1 Tall deciduous trees should be planted around roads, squares and outdoor parking lots as well as inside outdoor parking lots for shading. The shading ratio shall be no less than 40% for squares in residential areas, no less than 20% for squares around public buildings, no less than 30% for open parking spaces, and no less than 75% for sidewalk and bicycle lane;

　　2 Deciduous broad-leaf trees should be planted on the east, south and west facades of buildings; if conditioned, vertical greening may be designed for shading the building facades;

　　3 Wind break should be designed in the upwind direction of the prevailing wind direction in winter in combination with the analysis report on the wind environment of site, so as to effectively resist to the prevailing wind in winter; wind leading forest belts should be planted in positions with static wind to provide favorable conditions for the natural ventilation of buildings in summer.

12.3 Waterscape

12.3.1 The existing water bodies on site, such as lakes, rivers and west lands, should be fully reserved on the basis of meeting the planning and design requirements; ecological design is required for the renovation of water bodies.

12.3.2 For waterscape design, the water balance in the site shall be comprehensively considered in combination with the weather conditions, landforms, conditions of water sources, mode of rainwater utilization, requirements for regulation and storage of rainwater, etc., and the reasonable waterscape scale shall be determined in combination with the facilities for collecting rainwater, etc.

12.3.3 Artificial waterscape shall be not designed in the land where nontraditional water resources are unavailable.

12.3.4 The design of artificial waterscape shall pay attention to the impact of seasonal variation on the waterscape effect, and shall give full consideration to the effects of it in dry seasons.

12.3.5 The technical measures of filtration, circulation, purification, oxygenation, etc. shall be taken for artificial waterscape.

12.4 Sites

12.4.1 The rainwater runoff on ground surface and roof shall be reasonably planned, effective techniques shall be employed for the infiltration, retention, regulation, storage and reuse of rainwater, and the runoff volume of rainwater in the site shall be controlled without any pollution to the environment.

12.4.2 For design of outdoor roads and squares, the facilities of sunshade, wind and rain shelter, etc. shall be considered, and the pavement materials for outdoor hard ground shall be selected in compliance with the principles of smoothness, light color, wear and skid resistance, and water permeability. Wherein, the hard pavement materials should have the solar radiation absorption factor of 0.3~0.7, the permeable pavement ratio shall not be less than 70% of all pavements, and the permeable pavement base course shall be formed by permeable construction.

12.4.3 The barrier-free design of outdoor places shall meet the following requirements:

    1 It shall meet the requirements in GB 50763 *Codes for Accessibility Design*;

    2 Barrier-free facilities shall be designed following the principle of humanization under the premise of satisfying the functions;

    3 For the design of public parking lots, it shall be considered to arrange special parking spaces for the disabled in the nearest position to the main entrance of buildings.

12.4.4 For the design of outdoor parking lots, the factors of sunshade, noise reduction, visual effects, etc. shall be considered, and trees and shrubs can be planted; for the pavement of outdoor parking lots, ecological and environment-friendly materials with good water permeability shall be selected.

12.4.5 The design of gymnasiums and fitness facilities for residents, shall meet the requirements of GB 50180 *Code of Urban Residential Areas Planning & Design*, and the land area shall meet the requirements of *Urban Community Sports Facilities Construction Land*. It should also meet the following requirements:

    1 Gymnasiums for residents shall be built in a centralized mode in the place convenient to reach for people;

    2 Fitness facilities for residents shall be arranged in combination with green space, and special fitness facilities shall be considered for elderly people;

    3 Gymnasiums shall have good sunlight and ventilation, and shall be provided with seats for rest.

12.4.6 Play ground for children shall be built in the area with sufficient sunshine and good wind environment, it should be designed to be open for ensuring the good inter-visibility, and it shall keep a certain distance from the main roads and residential windows. In the play ground, safe facilities in suitable dimensions shall be selected, and cleaning tank special for children should be provided as well.

12.4.7 For the design of pavilions, sculptures, artistic devices and other street furniture, the functions of sunshade, wind shelter, sound-proof barrier, etc. should be considered.

12.4.8 For the design of street furniture, local materials, reusable materials, recycled materials and environment-friendly materials should be considered and selected in priority.

12.4.9 For the public facilities of outdoor heating plants or heat exchange station, transformer room, switching station, switch house for streetlamps, gas offtake station, high-pressure pump station, public toilet, refuse transfer station and pick-up point, bicycle park for residents, parking lots (garage) for residents, etc. , shield, fence or beautification measures should be designed under the premise of no influence on the function and warning.

12.5 Lighting

12.5.1 Besides the requirements in this section, the nightscape lighting design shall also meet the requirements in 11.3 of this standard.

12.5.2 The nightscape lighting design shall comply with the principles of safety and moderation, and shall be in accordance with the relevant requirements in JGJ/T 163 *Code for Lighting Design of Urban Nightscape* and DB11/T 388.1-4 *Technical Code of Urban Nightscape Lighting*.

12.5.3 The nightscape lighting design shall employ effective techniques for restricting light pollutions and shall meet the following requirements:

1 For nightscape lighting, the light shall be strictly controlled within the site, and the light spilling out of the site shall not exceed 15%;

2 The nightscape lighting facilities shall be strictly controlled to avoid producing interferential light to the residential buildings, apartments, hospital wards, etc. , and shall meet the requirements of 7.0.2 in JGJ/T 163-2008 *Code for Lighting Design of Urban Nightscape* and Section 5.2 in DB11/T 388.3-2006 *Technical Code of Urban Nightscape Lighting*;

3 The working time of nightscape lighting shall be set reasonably, and the internal transmitted lighting of nightscape lighting shall be partially or completely turned off in time;

4 For nightscape lighting of glass curtain walls, building facades made of the surface materials with the reflectivity factor less than 0.2, the combination of internal transmitted lighting with outline lighting should be used and flood lighting shall not be adopted; 5 The technique of shielding angle should be employed for light sources with the initial luminous flux exceeding 1000 lm.

12.5.4 For nightscape lighting, the light sources, lamps and accessories shall be selected in accordance with the requirements in Section 3.2 of JGJ/T 163-2008 *Code for Lighting Design of Urban Nightscape*. Besides satisfying the lighting function, the selection of lamps for nightscape lighting shall also pay attention to the landscaping effect in daytime.

12.5.5 Control in separation shall be performed for nightscape lighting of public buildings by normal days, festivals and major festivals.

12.5.6 For the nightscape lighting of building facades, the lighting power density shall meet the requirements for large cities specified in 6.2.2 of the current professional standard JGJ/T 163-2008 *Code for Lighting Design of Urban Nightscape*.

12.5.7 If conditioned, nightscape lighting facilities may be designed in integration with the facilities for photovoltaic power generation, wind power generation, etc.

13 Interior Decoration Design

13.1 General Requirements

13.1.1 Decoration design shall comply with the principles of high efficiency, health and moderation.

13.1.2 For newly built buildings, the interior decoration should be designed in integration with the buildings.

13.1.3 For the decoration design of existing buildings, the main structure shall not be destroyed, the performance of construction equipment shall not be affected, and the positions of electromechanical device terminals should not be changed.

13.1.4 The decoration design shall take the dismantlability of decoration materials, parts, facilities, etc. into account, and shall provide possibility through the decoration structure design. For office buildings, commercial buildings, flexible partition walls should be used in indoor space to reduce the waste of materials and the production of refuses in redecoration.

13.2 Design Requirements

13.2.1 Indoor decoration design shall not weaken the sound insulation of room envelope.

13.2.2 Indoor decoration design shall not affect indoor natural lighting, and the positions of external window, internal window, balcony, etc. should not have any shielding component except internal sun shading.

13.2.3 Indoor decoration design shall not weaken the thermal performance of the external envelope of buildings, and shall avoid producing thermal bridge at the same time.

13.2.4 For indoor decoration design, the indoor air quality shall be pre-evaluated, and the results of the pre-evaluation shall meet the requirements specified in Table 6.0.4 of GB 50325 - 2001 *Code for Indoor Environmental Pollution Control of Civil Building Engineering*.

13.2.5 For indoor decoration design, the safety of the waterproof positions shall be strictly ensured.

13.2.6 The lighting design for interior decoration shall provide the calculation sheet for energy saving of lighting, and shall meet the relevant requirements in Section 11.3.

13.2.7 For indoor decoration design, materials and structures hard for dust retention and easy to clean should be selected.

13.2.8 Measures such as green plants and reasonably designed waterscape should be considered for indoor decoration design for improving air quality and adjusting indoor humidity.

13.2.9 Industrialized and prefabricated complete products should be selected for indoor decoration design.

13.2.10 Indoor decoration design shall provide and reserve access holes for electromechanical equipment.

13.3 Selection of Decoration Materials

13.3.1 The contents of harmful substances of indoor finishing materials shall meet requirements of the current national standards. The indoor finishing materials shall be selected in strict accordance with the current requirements for the restricted and forbidden building materials and products promulgated by the state and Beijing area. The indoor finishing materials shall meet the requirements for the technical indicator system for material selection of green buildings in the national standard GB/T 50378 *Evaluation Standard for Green Building* and DB 11/T 825 *Evaluation Standard for Green Building* of Beijing.

13.3.2 Recycled materials and waste-reuse materials should be selected for interior decoration.

13.3.3 Durable materials, energy-storing materials, and functional materials capable of self-cleaning, removing formaldehyde, resisting bacteria, improving indoor air quality, etc. should be selected for interior decoration.

13.3.4 For the bamboo and wood materials in interior decoration, fast-growing wood and synthetic high strength composite materials should be selected. Native materials should be used as much as possible on the basis of ensuring the decorative effects.

14 Special Design Control

14.1 General Requirements

14.1.1 Prior to developing special design, the major design organization shall firstly conduct the feasibility study regarding the special scheme.

14.1.2 Prior to developing special design, the major design organization shall provide necessary design conditions and define the design requirements; on completion of the special design, the major design organization shall conduct complete checking for the special design.

14.1.3 The building curtain wall system, solar hot water system and solar photovoltaic power generation system shall be coordinated with buildings, and shall be able to ensure the safety.

14.1.4 Special projects must keep unified planning and simultaneous design with the major project.

14.1.5 Special design shall draft the corresponding operation control strategies and maintenance schemes based on the requirements for the major design.

14.2 Building Curtain Wall

14.2.1 The divisions of glass curtain walls shall be matched with the indoor special organization, but shall not obstruct the indoor functions and visual requirements of the design for main works.

14.2.2 For glass curtain walls, the effective ventilating area of the openable part shall meet the design requirements for main works.

14.2.3 The special design of curtain walls shall ensure to meet the requirements for the thermal performance of the design for main works. Heat engineering calculation shall be performed for the envelope part installed with building curtain walls, and corresponding techniques for thermal insulation shall be employed for the thermal bridge part.

14.2.4 Special design shall meet the requirements in the main design for sound insulation and noise reduction of glass curtain walls; the reflectivity, light transmittance, shading coefficient, air tightness, etc. of glass shall meet the design requirements for main works.

14.2.5 Unit curtain walls produced by industrialized production should be used as glass curtain walls, whereas back bolt type hanging stone curtain walls should be used as stone curtain walls.

14.2.6 For design of curtain wall, various materials such as stones, adhesives, heat-insulating rock wool, etc. shall be selected in accordance with the requirements for environmental protection.

14.3 Reclaimed water Treatment and Rainwater Recycling Systems

14.3.1 The process flow of reclaimed water treatment shall be determined after economic and technical comparison according to the water quality and water quantity of raw reclaimed water, together with the water quality, water quantity and operating requirement of reclaimed water, etc.

14.3.2 If the community has the landscape water system, the rainwater collection and utilization system shall be combined with the design of the landscape water system in the district; natural and ecological methods shall be adopted in priority to collect, treat, store, utilize or infiltration rainwater.

14.3.3 Under the premise of the water quality of reclaimed water, mature treatment technology and equipment with low energy consumption, high efficiency and easy maintenance shall be adopted.

1 When gray water or high grade gray water is used as the raw reclaimed water, the main process flow of physio-chemical treatment or the process flow combining biological treatment with physio-chemical treatment should be selected;

2 When the drainage containing fecal sewage is used as the raw reclaimed water, the process flow of treatment combining secondary biological treatment with physio-chemical treatments should be selected;

3 When the water after secondary treatment by sewage treatment station is used as the raw reclaimed water, the process flow of advanced treatment combining physio-chemical treatment with biochemical treatment should be selected.

14.3.4 The rainwater treatment technology shall be designed according to the relevant requirements of the current national standard GB 50400 *Engineering Technical Code for Rain Utilization in Building and Sub-district*.

14.3.5 The water quality of reclaimed water and reuse rainwater shall be determined according to the application, and shall meet the requirements of relevant national current standards and specifications.

14.3.6 The reclaimed water treatment facilities, rainwater collection and utilization system shall employ techniques for safety control and monitoring of water quality as well as water quantity, and shall not produce any adverse impact on human health and surrounding. It is strictly forbidden to lead any reclaimed water and reuse rainwater into the water supply system of domestic drinking water.

14.4 Solar Thermal and Photovoltaic Systems

14.4.1 The type of solar water heating system shall be selected according to the factors of building type, operating requirement, operating mode, installation conditions, etc., and shall meet the requirements for safety, applicability, economy and good-looking appearance.

14.4.2 The solar water heating system should make full use of the feed pressure.

14.4.3 The solar water heating system shall be safe and reliable, the built-in heating system must be equipped with a device capable of ensuring the safety in use, and the following technical measures shall be adopted: anti-freezing, anti-condensation, overheating prevention, lightning protection, anti-hail, wind resistance and seismic resistance, etc..

14.4.4 The thermal performance of solar heat collection system shall meet the requirements of the current national standards for solar products. The normal life cycle of the major components in the system, such as heat collector, water storage tank and bracket, etc. shall not be shorter than 15a.

14.4.5 For centralized solar water heating system, the heat collection system should be designed according to the solar guarantee rate of 50%~60%. Refer to Table B.0.2-3 in Appendix B of this standard for the design parameters.

14.4.6 The solar hot water system shall be equipped with an automatic control system which ensure the maximum utilization of solar energy.

14.4.7 The solar hot water system shall be provided with an auxiliary energy heating device. The type of the auxiliary energy heating device shall be selected according to the using characteristics of buildings, using amount of hot water, energy supply, maintenance management, sanitary antibacteria, etc., and shall meet the relevant provisions of the current national standard GB 50015 *Code for Design of Building Water Supply and Drainage*.

14.4.8 The centralized solar water heating system shall meter the using amount of the auxiliary energy heating device; water meters shall be installed on the water supply pipeline of solar hot water and the water feeding pipeline of hot water storage tank.

14.4.9 The design of photovoltaic system shall meet the relevant provisions of the current national standard JGJ 203 *Technical Code for Application of Solar Photovoltaic System of Civil Buildings*.

14.4.10 The lightning-proof and grounding techniques for the photovoltaic system and grid-connected interface equipment shall meet the relevant provisions of the current national standards SJ/T 11127 *Overvoltage Protection for Photovoltaic (PV) Power Generating Systems -Guide* and GB 50057 *Design Code for Protection of Structures Against Lightning*.

14.4.11 The grid-connected interface equipment shall meet the relevant provisions of the current national standard GB/T 19939 *Technical Requirements for Grid Connection of PV System*, and meet the following requirements as well:

 1 An isolation device shall be set between photovoltaic system and public grid;

 2 The grid-connected interface equipment shall own the functions of automatic detection and grid connection cutoff protection.

14.4.12 The solar photovoltaic power generation system shall be equipped with power metering devices for generated energy.

14.4.13 The solar photovoltaic power generation system should be equipped with the control system capable of realizing the data acquisition and remote transmission of real-time generated energy, accumulated generated energy, etc.

14.5 Heat Pump System

14.5.1 Ground-source heat pump system must be designed in accordance with the geological and hydrogeologic conditions of the site, mainly including lithology, the water temperature, water quality, water quantity and water level of underground water, perennial temperature and heat-transfer characteristics of soil. Refer to Appendix B.0.7 of this standard for the conditions of shallow geothermal energy.

14.5.2 The design of sewage source heat pump system must take grasping the conditions of the sewage resources in the project location as the premise, including the water quality, water quantity, water temperature, flow path and changing laws of the sewage available at present; meanwhile, the changes of the sewage resources in the future shall be objectively evaluated.

14.5.3 The design of ground-source heat pump system shall not destroy the natural eco-environment of the area where the project is located.

 1 For underground water source heat pump system, effective techniques for recharge shall be employed to ensure that all the underground water is recharged to the same aquifer, and not any pollution shall be produced to the underground water resources;

2 For soil source heat pump system, the heat balance shall be calculated between the heat extraction on the source side and the heat rejection, so as to avoid the continuous rising or falling of soil temperature caused by the imbalance between heat extraction and heat rejection.

14.5.4 Heat pump systems shall be provided with metering devices by items for heat supply and driving energy.

14.6 Ice Storage System

14.6.1 The use of ice storage system should comply with the following conditions:

1 The scale of building using the ice storage system should be larger than 30,000m^2;

2 The duration when the design hourly load is 70% lower than the peak load shall not be less than 7h;

3 For air-conditioning, the superposition of the peak time of design hourly load with the peak time of grid shall not be less than 3h;

4 For air-conditioning, the superposition of the valley time of design hourly load with the valley time of grid shall not be less than 7h.

14.6.2 Secondary refrigerant should select ethylene glycol, and the proportioning concentration shall be accurately determined according to the operating temperature.

14.6.3 Ice storage system shall be equipped with an electric water mixing valve to control the water supply temperature, so as to ensure the stability of water supply temperature.

14.6.4 For ice storage device, the changes of cold ratio shall not exceed 20%, and the fluctuation range of water temperature shall not be larger than $\pm 1°C$.

14.6.5 The load factor of refrigerating conditions shall not be less than 0.65 for refrigerators of dual operating conditions.

14.7 Building Intelligent Systems

14.7.1 In the design of intelligent system, the part related to construction equipment and systems shall fully realize all the professional original design intentions, and the optimal strategies for operation control shall be established according to the control technology and requirements for operating management of the construction equipment and systems for heating, ventilation, air-conditioning and cooling, water supply and sewerage, lighting, elevator, etc.

14.7.2 In order to realize the optimal control of construction equipment and systems, the intelligent system shall be designed with complete list of monitoring points, and shall have the corresponding function of operating management.

14.7.3 For large-scale public buildings, the intelligent system shall realize the functions of monitoring, control, metering, statistics, analysis, etc. for various energy and water systems and equipment, and should have the display function.

Appendix A Integrated Review Sheet

A. 0. 1 Integrated Review Sheet (Residential Building, Scheme Stage)

Table A. 0. 1 Integrated Review Sheet (Residential Building, Scheme Stage)

01 Basic Information   Filled by:

Project name					
location					
Floor area	m²	Total land area		hm²	Floor area ratio
Building type	(Commercial housing, indemnificatory housing)		Building height		
Owner			Contact person and way		
Design organization			Contact person and way		
Consulting organization			Contact person and way		
Design starting and ending time			Design stage		
Project progress			Application time planned		
Application level planned					

02 Architecture Specialty (the part in gray background is optional)   Filled by:

		Shape factor		Window-wall ratio	East	South	West	North	Roof transparent area ratio	
Basic design information of building envelope		Construction of non-transparent building envelope							Heat transfer coefficient	
		Construction of transparent building envelope							Heat transfer coefficient	
		Energy saving rate of building envelope								
Implementation of planning indicators		P2 Residential land area per capita					P14 Rainwater runoff discharge			
		P3 Underground building floor area ratio					P15 Ratio of sunken green space			
		P7 Ratio of barrier-free residences					P16 Permeable pavement rate			
		P9 Outdoor parking ratio					P17 Green Ratio			
		P10 Energy consumption per unit floor area					P18 Roof greening ratio			
		P11 Contribution rate of renewable energy					P19 Woodlot ratio			
		P12 Average daily rated water consumption					P20 Native plant index			

continue A.0.1

Implementation of building indicators			D7 Utilization rate of waste-reuse materials	
	D2 Distance from building entrance/exit to public transport station		D8 Utilization rate of recycled materials	
	D4 Area ratio of active exterior-shading		D9 Indoor noise compliance ratio of main function space	
	D5 Cost ratio of purely decorative components			
Brief description of green building scheme: (can be attached below)				

03 Structure (the part in grey background is optional)　　　　　　　　　　　Filled by:

Structure design life		Structure system		
Seismic fortification intensity		Whether the industrialized building system is used	Yes No	Which system
D10 Ratio of high strength steel bars		D11 Ratio of high strength concrete		
D12 Ratio of high performance steel				
Brief description of structure design (can be attached below)				

04 Water Supply and Sewerage　　　　　　　　　　　　　　　　　　　　Filled by:

Whether nontraditional water source is used for greening water	Yes No	D13 Utilization rate of water-saving devices and equipments	
Utilization mode of nontraditional water source	Municipal reclaimed water　Self-built reclaimed water		
Rainwater utilization mode	Local infiltration　Collection and utilization		
D14 Utilization rate of nontraditional water source		D15 Utilization rate of water-saving irrigation for green spaces	
Whether the solar water heating system is used	Yes No		
Brief description of water supply and sewerage design (can be attached below)			

05 Heating, Ventilation, Air-conditioning and Cooling (the part in grey background is optional)　　Filled by:

Heating mode		Air-conditioning mode	
D16 Summated refrigerating coefficient of performance (SCOP) of water chilling (heat pump) unit with centralized cooling source	(If any)	D17 Coefficient of performance (COP) of water chilling (heat pump) unit with centralized cooling source	(If any)
D18 System transfer and distribution efficiency			
Brief description of design (can be attached below)			

continue A.0.1

06 Electricity (the part in grey background is optional)　　　　　　　　　　　　　　　　　　　　Filled by:

D19 Lighting power density of main spaces		D20 Target energy efficiency of transformer	
Brief description of design (can be attached below)			

07 Landscape　　　　　　　　　　　　　　　　　　　　　　　　　　　　　　　　　　　　　Filled by:

D21 Lighting power density of building facade nightscape		D24 Shading ratio of sidewalk and bicycle road	
D22 Solar radiation absorption factor of rigid pavement (average value)		D25 Number of arbors/100m² green space	
D23 Shading ratio of outdoor parking spaces		D26 species of woody plants	
Brief description of design: (can be attached below)			

08 Interior Decoration　　　　　　　　　　　　　　　　　　　　　　　　　　　　　　　　Filled by:

D27　Ratio of construction and decoration integration	
Brief description of design:	

09 Notes　　　　　　　　　　　　　　　　　　　　　　　　　　　　　　　　　　　　　　Filled by:

(Brief introduction of project features)

Description:

1. The design organization shall fill in truthfully in correspondence to different design stages according to the actual design situation.
2. Please refer to Chapter 4 "Index System" in this standard for the index reference limits and corresponding calculation methods.
3. The parts in grey background are optional contents in scheme stage.
4. The description part can be attached below if excessive.

A.0.2 Integrated Review Sheet (Public Building, Scheme stage)

Table A.0.2　Integrated Review Sheet (Public Building, Scheme stage)

01 Basic Information　　　　　　　　　　　　　　　　　　　　　　　　　　　　　　　　Filled by:

Project name					
Location					
Floor area	m²	Total land area		hm²	Floor area ratio
Building type	(Commercial housing, indemnificatory housing)		Building height		
Owner			Contact person and way		
Design organization			Contact person and way		
Consulting organization			Contact person and way		
Design starting and ending time			Design stage		
Project progress			Application time planned		
Application level planned					

continue A.0.2

02 Architecture Specialty (the part in gray background is optional)   Filled by:

		Shape factor		Window-wall ratio	East	South	West	North	Roof transparent area ratio	
Basic design information of building envelope		Construction of non-transparent building envelope							Heat transfer coefficient	
		Construction of transparent building envelope							Heat transfer coefficient	
		Energy saving rate								
Implementation of planning indicators		P3 Underground building floor area ratio				P15 Ratio of sunken green space				
		P7 Ratio of barrier-free guest rooms (for hotels)				P16 Permeable pavement rate				
		P9 Outdoor parking ratio				P17 Green Ratio				
		P10 Energy consumption per unit floor area				P18 Roof greening ratio				
		P11 Contribution rate of renewable energy				P19 Woodlot ratio				
		P12 Average daily rated water consumption				P20 Native plant index				
		P14 Rainwater runoff discharge								
Implementation of building indicators		D1 Compliance ratio of barrier-free design				D7 Utilization rate of waste-reuse materials				
		D2 Distance from building entrance/exit to public transport stations				D8 Utilization rate of recycled materials				
		D5 Cost ratio of purely decorative components				D9 Indoor noise compliance ratio of main function space				
		D6 Area ratio of space enclosed by recycled partitions								

Brief description of green building scheme: (can be attached below)

03 Structure (the part in gray background is optional)   Filled by:

Structure design life		Structure system		
Seismic fortification intensity		Whether the industrialized building system is used	Yes No	Which system
D10 Ratio of high strength steel bars		D11 Ratio of high strength concrete		
D12 Ratio of high performance steel				
Brief description of scheme				

· 63 ·

continue A.0.2

04 Water Supply and Sewerage   Filled by:

Whether nontraditional water source is used for greening	Yes No	D13 Utilization rate of water-saving devices and equipments	
Utilization mode of nontraditional water source		Municipal reclaimed water Self-built reclaimed water	
Rainwater utilization mode		Local infiltration Collection and utilization	
Whether the solar water heating system is used		Yes No	
D14 Utilization rate of nontraditional water source		D15 Utilization rate of water-saving irrigation for green spaces	
Brief description of scheme			

05 Heating, Ventilation, Air-conditioning and Cooling (the part in gray background is optional)   Filled by:

Heating mode		Air-conditioning mode	
D16 Summated refrigerating coefficient of performance (SCOP) of water chilling (heat pump) unit with centralized cooling source		D17 Coefficient of performance (COP) of water chilling (heat pump) unit with centralized cooling source	
D18 System transfer and distribution efficiency			
Brief description of scheme			

06 Electricity (the part in grey background is optional)   Filled by:

D19 Lighting power density of main spaces		D20 Target energy efficiency of transformer	
Brief description of scheme			

07 Landscape   Filled by:

D21 Lighting power density of building facade nightscape		D22 Solar radiation absorption factor of rigid pavement (average value)	
D23 Shading ratio of open parking spaces		D24 Shading ratio of sidewalk and bicycle road	
D25 Number of arbors/100m² green space		D26 species of woody plants	
Brief description of scheme			

08 Notes   Filled by:

(Brief introduction of project features)

Description:

1. The design organization shall fill in truthfully in correspondence to different design stages according to the actual design situation.
2. Please refer to Chapter 4 "Index System" in this standard for the index reference limits and corresponding calculation methods.
3. The parts in grey background are optional contents in scheme stage.
4. The description part can be attached below if excessive.

A.0.3 Integrated Review Sheet (Residential Building, Construction Drawing Stage)

Table A.0.3 Integrated Review Sheet (Residential Building, Construction Drawing Stage)

01 Basic Information   Filled by:

Project name					
Location					
Floor area	m²	Total land area		hm²	Floor area ratio
Building type	(Commercial housing, indemnificatory housing)		Building height		
Owner			Contact person and way		
Design organization			Contact person and way		
Consulting organization			Contact person and way		
Design starting and ending time			Design stage		
Project progress			Application time planned		
Application level planned					

02 Architecture Specialty   Filled by:

		Shape factor		Window-wall ratio	East	South	West	North	Roof transparent area ratio	
Basic design information of building envelope		Construction of non-transparent building envelope							Heat transfer coefficient	
		Construction of transparent building envelope							Heat transfer coefficient	
		Energy saving rate of building envelope								
Implementation of planning indicators		P2 Residential land area per capita				P14 Rainwater runoff discharge				
		P3 Underground building floor area ratio				P15 Ratio of sunken green space				
		P7 Ratio of accessible residences				P16 Permeable pavement rate				
		P9 Outdoor parking ratio				P17 Green Ratio				
		P10 Energy consumption per unit floor area				P18 Roof greening ratio				
		P11 Contribution rate of renewable energy				P19 Woodlot ratio				
		P12 Average daily rated water consumption				P20 Native plant index				
Implementation of building indicators		D1 Compliance ratio of accessible design				D7 Utilization rate of waste-reuse materials				
		D2 Distance from building entrance/exit to public transport stations				D8 Utilization rate of recycled materials				
		D4 Area ratio of active exterior-shading				D9 Indoor noise compliance ratio of main function space				
		D5 Cost ratio of purely decorative components								

continue A.0.3

Energy consumption per unit area of building sub-items	Heating		Air-conditioning		Lighting	
	Household appliances		Cooking		Domestic hot water	
	Elevator		Misc			

03 Structure   Filled by:

Structure design life		Structure system		
Seismic fortification intensity		Whether the industrialized building system is used	Yes No	Which system
D10 Ratio of high strength steel bars		D11 Ratio of high strength concrete		
D12 Ratio of high performance steel				

04 Water Supply and Sewerage   Filled by:

Utilization range of nontraditional water source	Indoor toilet flushing Outdoor greening Misc	D13 Utilization rate of water-saving devices and equipments	
Rainwater utilization		Integrated runoff coefficient	
		Collection and utilization scale	
Solar hot water utilization scale	Solar hot water generation per household		
	Solar collector area per household		
D14 Utilization rate of nontraditional water source		D15 Utilization rate of water-saving irrigation for green spaces	

05 Heating, Ventilation, Air-conditioning and Cooling   Filled by:

Heating mode		Air-conditioning mode	
D16 Summated refrigerating coefficient of performance (SCOP) of water chilling (heat pump) unit with centralized cooling source	(If any)	D17 Coefficient of performance (COP) of water chilling (heat pump) unit with centralized cooling source	(If any)
D18 System transfer and distribution efficiency			

06 Electricity   Filled by:

D19 Lighting power density of main spaces		D20 Target energy efficiency of transformer	

07 Landscape   Filled by:

D21 Lighting power density of building facade nightscape		D22 Solar radiation absorption factor of rigid pavement (average value)	
D23 Shading ratio of open parking spaces		D24 Shading ratio of sidewalk and bicycle road	
D25 Number of arbors/100m² green space		D26 Kinds of woody plants	

08 Interior Decoration continue A.0.3 Filled by:

| D27 Ratio of construction and decoration integration | |

09 Notes Filled by:

| (Brief introduction of project features) |

Description:
1. The design organization shall fill in truthfully in correspondence to different design stages according to the actual design situation.
2. Please refer to Chapter 4 "Index System" in this standard for the index reference limits and corresponding calculation methods.
3. The parts in gray background are optional contents in scheme stage.
4. The description part can be attached below if excessive.

A.0.4 Integrated Review Sheet (Public Building Construction Drawing Stage)

Table A.0.4 Integrated Review Sheet (Public Building Construction Drawing Stage)

01 Basic Information Filled by:

Project name					
Location					
Floor area	m²	Total land area		hm²	Floor area ratio
Building type	(Commercial housing, indemnificatory housing)		Building height		
Owner			Contact person and way		
Design organization			Contact person and way		
Consulting organization			Contact person and way		
Design starting and ending time			Design stage		
Project progress			Application time planned		
Application level planned					

02 Architecture Specialty Filled by:

	Shape factor		Window-wall ratio	East	South	West	North	Roof transparent area ratio	
Basic design information of building envelope	Construction of non-transparent building envelope							Heat transfer coefficient	
	Construction of transparent building envelope							Heat transfer coefficient	
	Energy saving rate								
Implementation of planning indicators	P3 Underground building floor area ratio				P15 Ratio of sunken green space				
	P7 Ratio of accessible residences (for hotels)				P16 Permeable pavement rate				
	P9 Outdoor parking ratio				P17 Green Ratio				

continue A. 0. 4

Implementation of planning indicators	P10 Energy consumption per unit floor area		P18 Roof greening ratio	
	P11 Contribution rate of renewable energy		P19 Woodlot ratio	
	P12 Average daily rated water consumption		P20 Native plant indicator	
	P14 Rainwater runoff discharge			
Implementation of building indicators	D1 Compliance ratio of accessible design		D7 Utilization rate of waste-reuse materials	
	D2 Distance from building entrance/exit to public transport stations		D8 Utilization rate of recycled materials	
	D5 Cost ratio of purely decorative components		D9 Compliance ratio of indoor noise of main function space	
	D6 Area ratio of space enclosed by recycled partitions			

Energy consumption per unit area of building sub-items	Air-conditioning cooling and heating loads				Lighting	
	Equipment		Elevator		Domestic hot water	
	Misc					

03 Structure — Filled by:

Structure design life		Structure system		
Seismic fortification intensity		Whether the industrialized building system is used	Yes No	Which system
D10 Ratio of high strength steel bars		D11 Ratio of high strength concrete		
D12 Ratio of high performance steel				

04 Water Supply and Sewerage — Filled by:

Utilization range of nontraditional water source	Indoor toilet flushing Outdoor greening Misc	D13 Utilization rate of water-saving devices and equipments	
Rainwater utilization	Integrated runoff coefficient		
	Collection and utilization scale		
Solar hot water utilization scale	Solar hot water generation per household		
	Solar collector area per household		
D14 Utilization rate of nontraditional water source		D15 Utilization rate of water-saving irrigation for green spaces	

05 Heating, Ventilation, Air-conditioning and Cooling — Filled by:

Heating mode		Air-conditioning mode	
D16 Summated refrigerating coefficient of performance (SCOP) of water chilling (heat pump) unit with centralized cooling source		D17 Coefficient of performance (COP) of water chilling (heat pump) unit with centralized cooling source	
D18 System transfer and distribution efficiency			

06 Electricity   continue A.0.4   Filled by:

D19 Lighting power density of main spaces		D20 Target energy efficiency of transformer	

07 Landscape   Filled by:

D21 Lighting power density of building facade nightscape		D22 Solar radiation absorption factor of rigid pavement (average value)	
D23 Shading ratio of open parking spaces		D24 Shading ratio of sidewalk and bicycle road	
D25 Number of trees/100m² green space		D26 Kinds of woody plants	

08 Notes   Filled by:

(Brief introduction of project features)

Description:

1. The design organization shall fill in truthfully in correspondence to different design stages according to the actual design situation.
2. Please refer to Chapter 4 "Index System" in this standard for the index reference limits and corresponding calculation methods.
3. The parts in gray background are optional contents in scheme stage.
4. The description part can be attached below if excessive.

Appendix B Beijing Design Data Collection

B. 0. 1 Outdoor Meteorological Parameters

The outdoor meteorological parameters may refer to Table B. 0. 1. The time span for the meteorological statistics is January 1, 1971 ~ December 31, 2000.

Table B. 0. 1 List of Meteorological Parameters

	Site	Beijing
	Station Name & No.	Beijing
		54511
Station Information	North Latitude	39°48'
	East Longitude	116°28'
	Elevation/m	31.3
	Statistic Years	1971-2000
Outdoor Design Temperature and Humidity	Mean Annual Temperature /℃	12.3
	Outdoor Design Temperature for Heating /℃	−7.6
	Outdoor Design Temperature for Ventilation in Winter /℃	−3.6
	Outdoor Design Temperature for Air-conditioning in Winter /℃	−9.9
	Outdoor Design Relative Humidity for Air-conditioning in Winter /%	44
	Outdoor Design Dry-bulb Temperature for Air-conditioning in Summer /℃	33.5
	Outdoor Design Wet-bulb Temperature for Air-conditioning in Summer /℃	26.4
	Outdoor Design Temperature for Ventilation in Summer /℃	29.7
	Outdoor Design Relative Humidity for Air-conditioning in Summer /%	61
	Daily Mean Outdoor Temperature for Air-conditioning in Summer for /℃	29.6
Wind Direction, Speed and Frequency	Mean Outdoor Wind Speed in summer /(m/s)	2.1
	Dominant Wind Direction in Summer	C SW
	Frequency of Dominant Wind Direction in Summer /%	18 10
	Mean Speed of Dominant Wind Direction in Summer /(m/s)	3.0
	Mean Outdoor Wind Speed in Winter /(m/s)	2.6
	Dominant Wind Direction in Winter	C N
	Frequency of Dominant Wind Direction in Winter /%	19 12
	Mean Speed of Dominant Wind Direction in Winter /(m/s)	4.7
	Annual Dominant Wind Direction	C SW
	Frequency of Annual Dominant Wind Direction /%	17 10

continue B.0.1

Sunshine Percentage in Winter /%		64
Maximum Depth of Frozen Ground/cm		66
Atmospheric Pressure	Outdoor Atmospheric Pressure in Winter /hPa	1021.7
	Outdoor Atmospheric Pressure in Summer /hPa	1000.2
Number of Days and Mean temperature for Design during Heating Period	Number of Days with the Daily Mean Temperature $\leqslant+5°C$	123
	Start and End Dates for Days with the Daily Mean Temperature $\leqslant+5°C$	11.12~03.14
	Mean Temperature During the Period with Mean Temperature $\leqslant+5°C/°C$	−0.7
	Number of Days with the Daily Mean Temperature $\leqslant+8°C$	144
	Start and End Dates for Days with the Daily Mean Temperature $\leqslant+8°C$	11.04~03.27
	Mean Temperature During the Period with Mean Temperature $\leqslant+8°C/°C$	0.3
Extreme Maximum Temperature /°C		41.9
Extreme Minimum Temperature /°C		−18.3

B.0.2 Hourly Meteorological Parameters for Simulation

The hourly meteorological parameters may be selected from the data of Beijing in *Meteorological Data Set Special for Analysis of Building Thermal Environment of China* which is collaborated by the Meteorological Data Room, the National Meteorological Information Center of China Meteorological Administration and the Department of Building Science & Technology, Tsinghua University. The main urban data of Beijing may be directly employed for the meteorological data concerning Pinggu, Shunyi, Haidian, etc. and the meteorological data of temperature, humidity, solar radiation, etc. The representative data of the local weather stations in the last 10 years shall be employed as much as possible for the meteorological data of wind direction, wind speed, precipitation, etc. Or the hourly meteorological data provided by the relevant weather departments shall be used. The data in Table B.0.2-1 may be employed for the statistics of wind direction and wind speed, and the data in Table B.0.2-2 may be employed as the data of horizontal solar radiation.

The data in Table B.0.2-3 may be employed for the monthly mean total daily solar radiation and annual mean daily radiation on the inclined surface. The data in Table B.0.2-3 are originated from the (*Atlas of National Building Standard Design: Selection and Installation of Solar Centralized Hot Water System*) (Atlas No.: 06SS128). The inclination angle of the inclined surface equals to the latitude of Beijing as 39°48'.

Table B.0.2-1 Charts of Wind Direction and Wind Speed for simulation

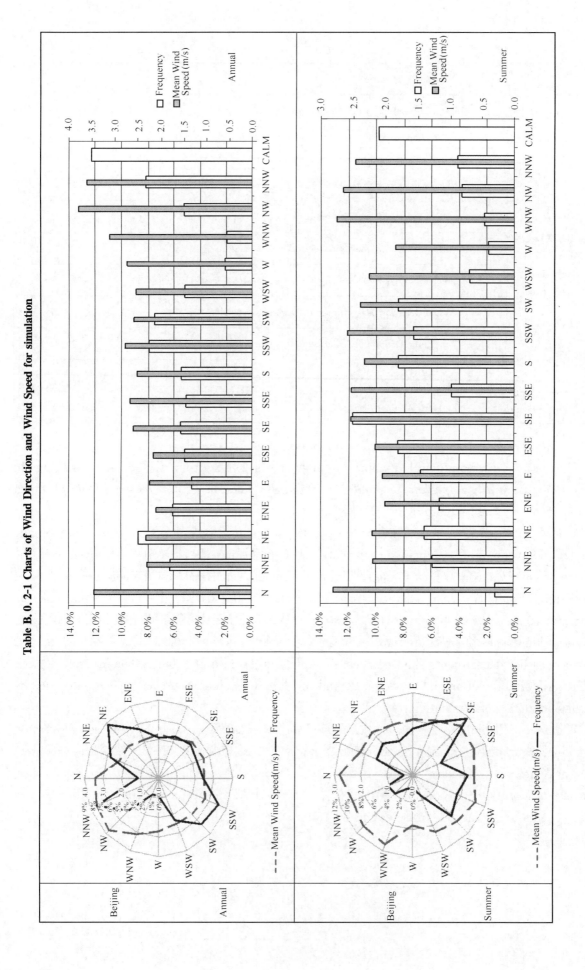

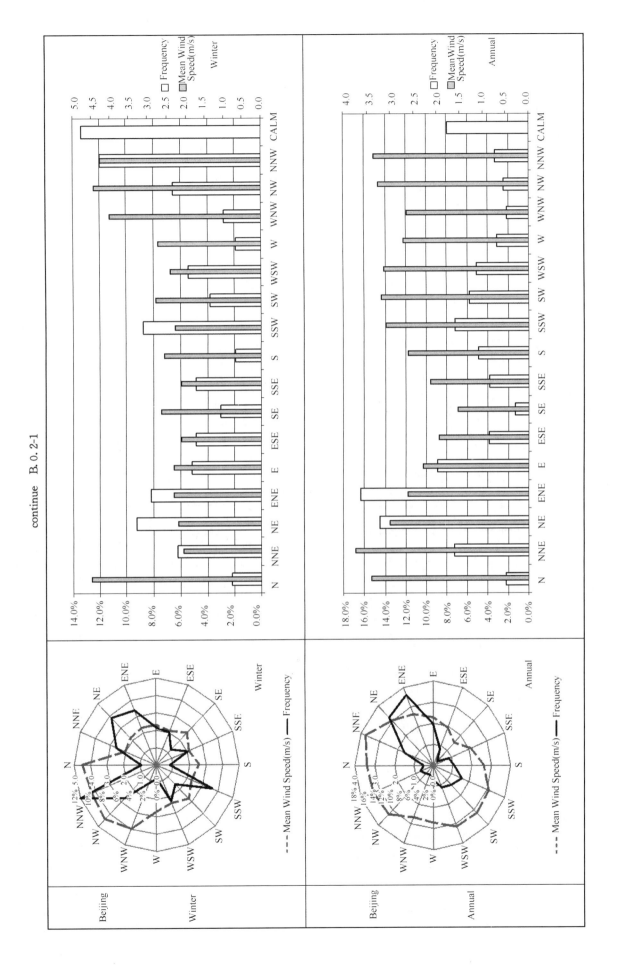

continue B.0.2-1

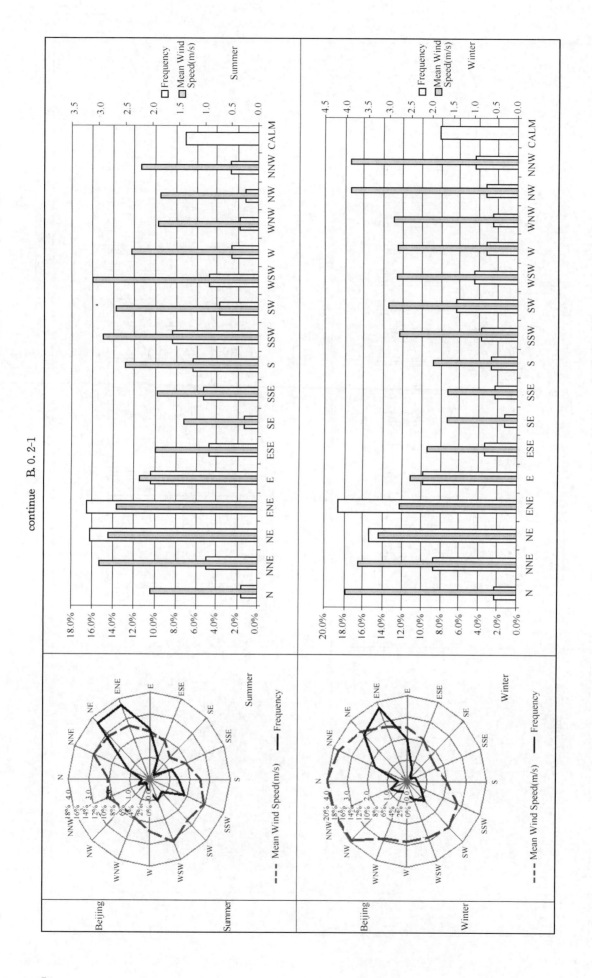

continue B.0.2-1

Table B. 0. 2-2 Chart of Monthly Total Radiation on Horizontal Plane for Simulation

	1	2	3	4	5	6	7	8	9	10	11	12
Monthly Total Direct Radiation (MJ/m^2)	180.59	234.85	259.20	334.10	460.10	399.35	313.39	253.08	222.01	188.63	162.93	123.55
Monthly Total Diffuse Radiation (MJ/m^2)	80.96	106.41	191.24	219.55	152.63	167.76	206.00	253.40	186.15	155.90	93.87	82.13

Table B. 0. 2-3 Monthly Mean Daily Average Radiation and Annual Mean Daily Average Radiation on Inclined Surface

City	January	February	March	April	May	June	July	August	September	October	November	December
Monthly Mean Daily Average Radiation (MJ/m$^2 \cdot$ d)	15.08	17.14	19.16	18.71	20.18	18.67	16.22	16.43	18.69	17.51	15.11	13.71
Annual Mean Daily Average Radiation (MJ/m$^2 \cdot$ d)	17.21											

B. 0. 3 Precipitation Conditions for Design

According to the precipitation data (2000~2010) provided by Beijing Meteorological Bureau, the precipitation related parameters for the main urban area of Beijing (observatory), Miyun, Pinggu, Yanqing, Mentougou and Huairou may refer to the data in Table B. 0. 3-1 and Table B. 0. 3-2.

Table B. 0. 3-1 Statistics of Annual Mean Precipitation and Mean Precipitation for Years in Beijing

	Observatory	Miyun	Pinggu	Yanqing	Mentougou	Huairou
Mean Precipitation for Years (mm)	485.1	625.5	611.3	465.2	591.1	582.4
Mean Rainy Days for Years	72.3	79.1	75.8	78.6	77.7	80
Ratio of precipitation above 2mm to total precipitation	95.2%	96.1%	96.3%	94.2%	95.5%	95.8%
Ratio of precipitation above 4mm to total precipitation	90.0%	91.6%	90.9%	86.5%	90.4%	89.9%

Table B. 0. 3-2 Monthly Mean Precipitation and Monthly Mean Rainy Days for Years

		January	February	March	April	May	June	July	August	September	October	November	December
Observatory	Monthly Mean Precipitation for Years (mm)	4.7	5.2	13.3	25.4	40.0	76.1	116.2	100.5	53.9	36.3	11.3	2.1
	Monthly Mean Rainy Days for Years	2.8	3.0	3.8	5.6	7.4	10.4	11.8	10.7	6.9	5.6	2.1	2.2
Miyun	Monthly Mean Precipitation for Years (mm)	4.1	4.1	12.5	22.2	53.6	105.5	146.2	156.7	67.8	41.5	9.0	2.3
	Monthly Mean Rainy Days for Years	2.1	2.2	3.7	5.9	8.4	11.9	13.3	12.3	7.2	7.6	2.2	2.3
Pinggu	Monthly Mean Precipitation for Years (mm)	4.0	4.5	12.8	25.1	50.4	116.9	147.5	137.3	58.9	38.8	11.8	3.1
	Monthly Mean Rainy Days for Years	2.5	2.3	3.5	6.0	7.7	10.8	13.5	10.7	7.6	6.6	2.2	2.4
Yanqing	Monthly Mean Precipitation for Years (mm)	3.6	2.7	12.4	21.8	48.7	78.0	116.4	80.2	64.4	27.1	7.3	2.6
	Monthly Mean Rainy Days for Years	2.1	2.4	3.6	6.6	8.8	12.3	11.7	11.6	8.7	6.8	2.0	2.0
Mentougou	Monthly Mean Precipitation for Years (mm)	4.0	4.8	13.8	23.9	46.3	96.8	172.6	129.9	55.6	29.2	11.1	3.0
	Monthly Mean Rainy Days for Years	2.4	2.6	3.3	5.7	7.8	12.1	13.9	11.5	7.9	6.3	2.2	2.0
Huairou	Monthly Mean Precipitation for Years (mm)	5.8	4.9	14.5	24.7	49.6	85.9	140.4	142.5	66.1	36.0	8.7	3.2
	Monthly Mean Rainy Days for Years	2.0	2.3	3.8	5.6	7.8	11.8	14.8	12.2	8.2	7.1	2.2	2.2

B. 0. 4 Material Resources

In order to ensure the quality of construction works, further improve the service functions of buildings, save the consumption of energy and other resources in the process of building construction and operation, protect environment and accelerate the healthy development of the building material industry, Beijing Municipal Commission of Housing and Urban-Rural Development and Beijing Municipal Commission of Urban Planning jointly promulgated the 2010 *Promoted, Restricted and Banned Product Catalog of Building Materials in Beijing* ("2010 Catalog" for short) on May 31, 2010. During the material selection for green building, the building materials and products promoted in Beijing shall be selected in priority so as to facilitate the popularization and application of new materials, new techniques, new equipment and new technology in Beijing, main building mater al mag refer to the data in Table B. 0. 4. In order to ensure the timeliness of the provision, all shall be subject to the latest and current catalog of promoted materials in Beijing. During the material selection for green building, all types of promoted materials or products shall be selected under the premise of meeting the requirements for the range of application in the catalog. Once the "2010 Catalog" is updated, the latest catalog shall be taken as the basis of selection.

Table B. 0. 4 List of Material Selection for Green Building

No.	Category	Name of Building Material	Promoted Range of Application	Reasons for Promotion
1	Concrete Material and Concrete Product	Recycled aggregate	Premixed concrete, premixed mortar and concrete products	The wastes formed in the process of demolishing buildings and structures are recycled, and it is in favor of resources saving and environmental protection

continue B.0.4

No.	Category	Name of Building Material	Promoted Range of Application	Reasons for Promotion
2	Wall Materials	B04/B05 aerated concrete building blocks and boards	Civil buildings	Light weight and good heat preservation
3		Heat reserving, structural and decorative integrated exterior wallboard	Civil buildings	Energy saving, fire protection and firm decorative layer
4		Gypsum hollow wallboard and building blocks	Filling materials for walls of buildings in frame structure	Light weight, sound insulation, energy saving, fire protection and making use of industrial wastes
5	Doors, Windows, Curtain Walls and Auxilliary Materials	High performance building external window with the heat transfer coefficient of $2.5W/(m^2 \cdot K)$ and below	Civil buildings	Capable of improving the energy saving level of buildings
6		Low-emissivity coated glass (low-E)	Exterior doors, windows and transparent curtain walls of civil buildings	Reducing the heat transfer coefficient of glass, and saving the energy consumption of buildings
7	Finishing Materials for Building Decoration	Light hanging panel made of architectural concrete	Decoration of interior and exterior walls for civil buildings	Good decorative effects, making use of waste residue, and high efficiency of construction
8		Ultrathin stone composite board	Decoration of interior and exterior walls for civil buildings	Saving high-quality natural stone resources, and reducing load of buildings
9		Flexible tapestry brick	Decoration of interior and exterior walls for civil buildings	Thin body, light weight, waterproofness, air permeability, good flexibility and convenient construction
10	Materials for Municipal and Road Construction	Water permeable brick (water permeability $\geqslant$ 30mL, average compression strength $\geqslant$ 40 MPa, average rupture strength $\geqslant$ 4 MPa)	Square, parking lot, sidewalk, bicycle road	Favorable for collecting rainwater to supplement urban underground water

B.0.5 Native Plants

Pay attention to preventing the invasion of alien species while selecting plants. Native plants have powerful adaptability, thus planting native plants may ensure the survival of the plants, reduce pest and disease damages and effectively lower the cost of maintenance. Refer to the list of native plants given in Appendix A~E of the local standard DB 11/T 211 - 2003 *Plant Materials for Urban Landscape Greening Tree Seedling* of Beijing (see Table B.0.5);

Table B. 0. 5 List of Common Plants in Beijing

Species	List of Plants
Evergreen Trees	*Abies holophylla*, *Picea koraiensis*, *Picea meyeri*, *Picea wilsonii*, *Cedrus deodara*, *Pinus tabulaeformis*, *Pinus bungeana*, *Pinus armandii*, *Platycladus orientalis*, *Sabina chinensis*, Xi'an Cypress, *Sabina Chinensis* 'Kaizuka', *Sabina komarovii*, *Ligustrum lucidum*
Deciduous Arbor	*Ginkgo biloba*, *Metasequoia glyptostroboides*, *Populus tomentosa*, *Salix matsudana*, *Salix babylonica*, *Salix matsudana* 'Umbraculifera', *Salix* × *aureo-pendula*, *Juglans regia*, *Pterocarya stenoptera*, *Quercus variabilis*, *Ulmus pumila*, *Ulmus pumila* cv. Pendula, *Zelkova schneiderlana*, *Celtis bungeana*, *Pteroceltis tatarinowii*, *Magnolia denudata*, *Magnolia biondii*, Saucer Magnolia, Hybrid Liriodendron, *Eucommia ulmoides*, *Platanus* × *acerifolia*, *Malus halliana*, Diamond Crabapple, Royal Crabapple, *Prunus cerasifera*, *Prunus serrulata*, *Prunus davidiana prunus sibirica*, *Albizia julibrissin*, Chinese Honeylocust, *Robinia pseudoacacia*, *Sophora japonica*, *Sophora japonica* 'Pendula' *Ailanthus altissima*, *Ailanthus altissima* 'Umbraculifera', *Euonymus bungeanus*, *Acer truncatum*, *Acer palmatum*, *Aesculus chinensis*, *Koelreuteria paniculata*, *Ziziphus jujula*, *Tilia mandshurica*, *Firmiana simplex*, *Elaeagnus angustifolia*, *Diospyros kaki*, *Diospyros lotus*, *Fraxinus velutina*, *Syringa pekinensis*, *Chionanthus retusus*, *Paulownia tomentosa*, *Catalpa ovata* and *Catalpa speciosa*
Evergreen Shrub	*Taxus cuspidata* var. umbraculifera, *Sabina procumbens*, *Sabina chinensis* 'Pfitzeriana', Meyer Juniper, Savin Juniper, CV. AUREA. NANA, *Cephalotaxus sinensis*, *Buxus sempervirens*, *Ilex cornuta*, *Euonymus japonicus*, *Euonymus japoniaus* Cuzhi, *Euonymus kiautschovicus* and Spanish Dagger
Deciduous Shrub	*Paeonia suffruticosa*, *Berberis thunbergii* 'Atropurpurea', *Chimonanthus praecox*, *Philadelphus pekinensis*, *Deutzia crenata*, *Ribes odoratum*, *Sorbaria kirilowii*, *Cotoneaster horizontalis*, *Cotoneaster multiflorus*, *Chaenomeles speciosa*, *Rosa chinensis*, *Rosa hybrida*, Ground Cover Roses, *Rosa xanthina*, *Kerria japonica* 'Pleniflora', *Rhodotypos scandens*, *Prunus* 'Duplex', *Prunus davidiana* 'Albo-plena' *Prunus persica* var. duplex, *Amygdalus persica* f. atropurpurea, *Amygdalus persica* 'Densa', *Amygdalus triloba* Plena, *Cerasus tomentosa*, *Prunus glandulosa*, *Prunus japonica*, *Prunus mume* var. tortuosa, *Prunus* × *blireana*, *Prunus* × *cirstena*, *Cercis chinensis*, *Indigofera kirilowii*, *Caragana sinica*, *Lespedeza fioribunda*, *Poncirus trifoliate*, *Cotinus coggygria*, *Cotinus coggygria atropurpureus*, *Hibiscus syriacus*, *Tamarix chinensis*, *Hippophae rhamnoides*, *Lagerstroemia indica*, single *Lagerstroemia indica*, safflower *Lagerstroemia indica*, white *Lagerstroemia indica*, flower *Punica granatum*, fruit *Punica granatum*, *Cornus alba*, *Adinandra mellettii*, *Cornus officinalis*, *Dendrobenthamia japonica*, *Forsythia suspense*, *Forsythia viridissima*, *Syringa oblata*, *Syringa oblatavar* 'Alba', *Syringa* × *persia*, *Syringa meyeri*, *Ligustrum quihoui*, *Ligustrum* × *vicaryi*, *Ligustrum obtusifolium*, *Jasminum nudiflorum*, *Clerodendrum trichotomum*, *Callicarpa dichotoma*, *Lycium barbarum*, *Weigela florida*, *Weigela florida* 'Red Prince', *Weigela coraeensis*, *Kolkwitzia amabilis*, *Abelia chinensis*, *Lonicera maackii*, *Lonicera tatarica*, *Sambucus racemosa* 'Aurea', *Viburnum sargentii* and *Viburnum farreri*
Evergreen Vine	*Euonymus fortunei* var. radicans, *Euonymus kiautschovicus* and *Hedera nepalensis*
Deciduous Vine	*Polygonum aubertii*, *Rosa multiflora*, *Rosa multiflora* 'Albo-plena', *Rosa banksiae*, Climbing Roses, *Wisteria Sinensis*, *Celastrus orbiculatus*, *Vitis amurensis*, *Parthenocissus tricuspidata*, *Parthenocissus quinquefolia*, *actinidia arguta*, *Actinidia chinensis*, *Campsis radicans* and *Lonicera* japonica.
Bamboos	*Phyllostachys propinqua*, *Phyllostachys nigra*, *Bambusa vulgaris* var. striata, *Phyllostachys aureosulcata* and *Indocalamus tessellatus*
Lawns and Ground Covers *	*Buchloe dactyloides*, *Zoysia sinica*, *Zoysia japonica*, *Festuca rubra*, *Festuca*, *Festuca arundinacea*, *Poa nemoralis*, *Poa pratensis*, *Poa Compressa*, *Poa annua*, Well-off Grass, *Agrostis stolonifera*, *Carex giraldiana*, *Carex rigescens*, *Trifolium repens*, *Iris tectorum*, *Hemerocallis fulva*, *Hosta plantaginea*, *Ophiopogor japonicus*, *Orychophragmus violaceus*, *Iris lactea*, *Viola philippica*, *Duchesnea indica* and *Taraxacum mongolicum*

Note: The items marked as " * " are cited from the list of common plants in Beijing of the national standard atlas of 03J012-2 *Landscape-Planting Design*.

B.0.6 Plants for Root Greening

The species recommended by DB11/T 281 *Code for Roof Greening* (see Table B.0.6) should be selected as partial Plants for roof greening.

TableB.0.6 List of Partial Plant Species Recommended for Roof Greening in Beijing

Species	List of Plants
Arbor	*Pinus armandii* *, *Pinus bungeana*, *Sabina chinensis* cv. Xi'an, *Sabina chinensis* 'Kaizuka', *Sabina chinensis*, *Sozhora japonica*, *Ginkgo biloba*, *Koelreuteria paniculata*, *Magnolia denudata* *, *Ulmus pumila* cv. Pendula, *Prunus cerasifera*, *Diospyros kaki*, Aesculus chinensis *, *Acer palmatum* *, *Cerasus yedoensis* *, Crabapple and *Crataegus Pinnatifida*
Shrub	*Sorbaria kirilouii*, *Euonymus japonicus* *, *Buxus microphylla*, *Yucca gloriosa*, *Ligustrum* × *vicaryi*, *Berberis thunbergii*, *Taxus cuspidata* *, *Forsythia suspense*, *Amygdalus triloba*, *Prunus* × *cirstena*, *Prunus japonica* *, *Amygdalus persica* 'Densa', *Syringa oblata*, *Kerria japonica* *, *Cornus alba*, *Rosa chinensis*, *Ranunculaceae* *, *Prunus persica* 'Duplex', *Jasminum nudiflorum*, *Lagerstroemia indica* *, *Lonicera maackii*, fruit *Punica granatum*, *Cercis chinensis* *, *Cotoneaster horizontalis*, *Primula poissonii*, *Cotinus coggygria*, *Weigela florida*, *Viburnum sargentii*, *Chionanthus retusus*, *Clerodendrum trichotomum*, *Hibiscus syriacus*, *Chimonantthus praecox* *, *Rosa xanthina*. and *Kolkwitzia amabilis*
Deciduous shrub	*Savin vulgalis*, *Euonymus japonicus*, *Taxus cuspidata*, *Buxus microphylla* var., *koreana Buxus microphylla*, *Sabina procumbens*
Ground Covers	*Hosta plantaginea*, *Iris lactea*, *Dianthus chinensis*, *Physostegia virginiana*, *Convallaria majalis*, *Matteuccia struthiopteris* *, *Trifolium repens*, *Euonymus fortunei* var. radicans, *Sabina vulgaris*, *Hibiscus moscheutos*, *Dendranthema morifolium*, *Paeonia lactiflora* *, *Iris tectorum*, *Hemerocallis fulva*, *Parthenocissus quinquefolia*, *Sedum*, *Hedera nepalensis* var. sinensis * and *Lonicerra tellmanniana* *

Notes: 1 The items marked as " * " refer to the plants requiring certain microclimate conditions in roof greening;
   2 They are cited from DB11/T 281—2005 *Code for Roof Greening* of Beijing.

B.0.7 Shallow Geothermal Resource

The resources of shallow geothermal energy are held in underground rock-soil, its storage, transportation, exploitation and utilization are all strictly restricted by regional geological and hydrogeological conditions, and there are greater differences in the manner and scale of resource utilization in different regions. The resource distribution of the shallow geothermal energy in plain terrain of Beijing is closely correlated with the hydrogeological conditions of the quaternary system. The lithological structure, thickness, granularity, aquifer thickness, water abundance, buried depth of water level, conditions of water supply and runoff, etc. of the quaternary system are the main factors restricting the occurrence, distribution and utilizability of shallow geothermal energy. According to Appendix 2 Geological Survey Report on Shallow Geothermal Energy Resource in plain terrain of Beijing in the *Implementation Plan for the Model Supporting Capability Building in Application of Renewable Energy Source Buildings in Beijing*:

For the plain terrain of Beijing, it is applicable to use underground water type ground-source heat pump for developing and utilizing shallow geothermal energy resources at the middle to upper part of alluvial-proluvial fan, whereas it is applicable to use develop and utilize shallow geothermal energy by means of buried pipes at the lower part of alluvial-proluvial fan and in the alluvial-proluvial plain. The zoning map is shown below:

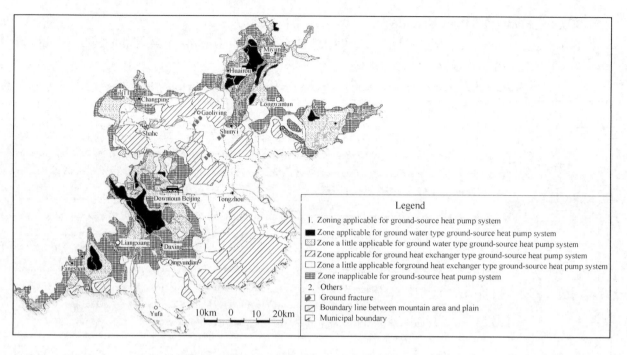

Figure B. 0. 7　Zoning Map of Shallow Geothermal Resource in Beijing

B. 0. 8　Basic Data for Carbon Emission Calculation

The data of building materials carbon emissions listed in Table B. 0. 8 is originated from BELES database, which developed by Department of Building Science and technology of Tsinghua University and represent the national level of production emission. The computation boundaries of building materials carbon emissions including the exploitation, processing and transportation of raw material, and the production of building material.

Table B. 0. 8　Basic Data for Carbon Emission Calculation

	Embodied Energy (MJ)	Carbon Emission (kg)	Data Source and Assumed Condition
Steel Product (kg)	22.7	3.3	Calculated according to the tonnage level of steel in North China; the data is originated from the statistic yearbook, open research articles and the achievements of project research during the "11th Five-year Plan"
Cement (kg)	5.3	1.1	Calculated according to the average distance of 529km for rail transport of non-metallic ores in 1990
Concrete (m^3)	3062.3	532.1	The proportioning is originated from the North China standard in 1990; non-metallic ores shall be calculated according to 529km for rail transport; cement shall be calculated according to 46km as the average tonnage mileage for highway transport in 1990
Aerated Concrete Building Block (m^3)	2221.3	332.9	Main raw materials and energy consumption shall refer to open research articles; water resources shall refer to Bill Laoson data; other materials shall be considered as recovered wastes and will not be calculated
Glass (kg)	20.1	1.4	Glass recovery may be considered as the effect of manual demolition of buildings. Glass density value is 2500kg/m^3
Expanded Polystyrene Panel (m^3)	3390.1	222.8	*Several Polymer Materials* only includes energy, raw materials and solid wastes; the energy consumption for the exploitation and transportation of petroleum raw materials is included in the energy content, and shall be calculated as highway transport in 2005

continue B.0.8

	Embodied Energy (MJ)	Carbon Emission (kg)	Data Source and Assumed Condition
Aluminum Product (kg)	43.7	10.9	Considered according to the recovery rate of 95%; electrolytic aluminum; the production of alumina is integrated with electrolytic process, the emission is originated from the coefficient manual, *Environmental Impact Assessment of Materials*, etc.
Architectural Ceramics (m²)	310.9	28.6	The haul distance of raw materials shall be calculated according to the rail transport of non-metallic ores

B.0.9 Survey on Average Level of Energy Consumption for Different Public Buildings

According to the 2007 *Statistical Summary Sheet for Energy Consumption of the Office Building of Beijing Municipal Government and Large-scale Public Buildings* and the *Annual Report on Development Research of Building Energy Saving in China* issued by Building Energy Saving Research Center, Tsinghua University, the survey on average level of energy consumption for different public buildings in Beijing is shown in Table B.0.9, including:

1) The energy consumption of public buildings refers to the statistics of energy consumption produced due to various activities in public buildings (the statistical value excludes the urban municipal heating), including that of air-conditioning, lighting, socket outlets, elevators, cooking, various service facilities and special functional equipment. The energy consumption of municipal heating and the energy consumption of the part of which the heating is provided by gas are not included in the statistics owing to the complicity in distinguishing the inside and outside of public buildings, in kWh/(m² · a);

2) It shall be classified according to building area and building functions. Buildings with the building area of no less than 20,000m² are taken as large-scale buildings, and those with the building area of less than 20,000m² are taken as common buildings.

Table B.0.9 Average Level of Energy Consumption for Different Public Buildings in Beijing

Building Type		Sample Size	Heating, Ventilation and Air-conditioning	Lighting	Indoor Equipment	Service	Others	Total
Large-scale Administration Building	Maximum	102	76.6	34	42.6	8.5	8.5	170.2
	Minimum		9.6	4.3	5.3	1.1	1.1	21.4
	Mean		36.8	14.7	18.4	1.5	2.2	73.6
Large-scale Commercial Office Building	Maximum	379	141.3	56.5	70.6	5.7	8.5	282.6
	Minimum		16.1	6.4	8.1	0.6	1	32.2
	Mean		60.7	27	27	13.5	6.7	134.9
Common Office Building	Maximum	32	29.2	13	13	6.5	3.2	64.9
	Minimum		3.5	6.4	3.2	4.8	3.2	21.1
	Mean		16.4	7.3	7.3	3.7	1.8	36.5
Large-scale Shop and Supermarket	Maximum	45	129.1	57.4	86.1	8.6	5.7	286.9
	Minimum		44.4	19.7	29.6	3	2	98.7
	Mean		61.6	27.4	41	4.1	2.7	136.8

continue B.0.9

Building Type		Sample Size	Heating, Ventilation and Air-conditioning	Lighting	Indoor Equipment	Service	Others	Total
Common Shop and Supermarket	Maximum	26	39.2	13.1	30.5	1.7	2.6	87.1
	Minimum		16.1	5.4	12.5	0.7	1.1	35.8
	Mean		33.7	11.2	26.2	1.5	2.2	74.8
Large-scale Hotel	Maximum	62	98.9	44	22	33	22	219.9
	Minimum		46.7	20.8	10.4	15.6	10.4	103.9
	Mean		71.8	31.9	16	23.9	16	159.6
Common Hotel	Maximum	25	70.7	19.3	19.3	12.9	6.4	128.6
	Minimum		13	3.6	3.6	2.4	1.2	23.8
	Mean		43.6	11.9	11.9	7.9	4	79.3
Large-scale School Building	Maximum	57	141.2	28.2	70.6	14.1	28.2	282.3
	Minimum		22.5	4.5	11.3	2.3	4.5	45.1
	Mean		44.7	8.9	22.3	4.5	8.9	89.3
Common School Building	Maximum	14	20.5	13.7	17.1	6.8	10.2	68.3
	Minimum		1.7	1.1	1.4	0.6	0.9	5.7
	Mean		6.6	4.4	5.5	2.2	3.3	22
Medical Building	Maximum	12	97.5	27.9	41.8	13.9	97.5	278.6
	Minimum		24.4	7	10.4	3.5	24.4	69.7
	Mean		48.3	13.8	20.7	6.9	48.3	138

Appendix C Boundary Conditions for Simulation

C.0.1 Outdoor Wind Environment Simulation

Simulation target:

The simulation of wind environment may be used for guiding the reasonable layout of building groups in building planning and design, optimizing the natural ventilation of site in summer and avoiding the unfavorable impact of the prevailing wind direction in winter. In the practical works, the boundary conditions shall be reasonably determined by using reliable computer simulation program, the wind environment of buildings shall be simulated based on the typical wind direction and wind speed, and the simulation shall also meet the following requirements:

1 For the pedestrian area around buildings, the wind speed at 1.5m high shall be less than 5m/s;

2 The magnification coefficient of wind speed shall be less than 2.

Input conditions[①]: It is recommended to refer to the research achievements by the European Cooperation in the Field of Scientific and Technical Research (COST) and the wind engineering research team of the Architectural Institute of Japan (AIJ) for simulation, so as to ensure the accuracy of the simulation results. In this standard, the research achievements by the Architectural Institute of Japan (AIJ) are employed.

In order to ensure the accuracy of the simulation results, the specific requirements are as follows:

1 Computational domain: The coverage area of building shall be 3% smaller than the area of the entire computational domain; the horizontal computational domain refers to the range taking the target building as the center and taking $5H$ as the radius. The computational domain above building shall be larger than $3H$; H is the height of the main body of building;

2 Model reproduced area: It shall be reproduced according to the maximum detail requirements within the range of boundary H of the target building;

3 Gridding: For the pedestrian height area on each side of building, the height of 1.5m or 2m shall be divided into 10 grids or more; the major observational area shall be in the third or the higher grid above the ground;

4 Boundary conditions for inlet condition: The distribution U (gradient wind) of inlet wind shall be given for simulation calculation, the k and ε for simulation may also be defined by distributive parameters under possible conditions;

$$U(z) = U_s \left(\frac{z}{z_s}\right)^\alpha \tag{C.0.1-1}$$

$$I(z) = \frac{\sigma_u(z)}{U(z)} = 0.1 \left(\frac{z}{z_G}\right)^{(-\alpha-0.05)} \tag{C.0.1-2}$$

[①] The research achievements by the Architectural Institute of Japan (AIJ).

$$\frac{\sigma_u^2(z)+\sigma_v^2(z)+\sigma_w^2(z)}{2} \cong \sigma_u^2(z) = (I(z)U(z))^2 \qquad \text{(C. 0. 1-3)}$$

$$\varepsilon(z) \cong P_k(z) \cong -\overline{uw}(z)\frac{\mathrm{d}U(z)}{\mathrm{d}z}$$

$$\cong C_t^{1/2} k(z) \frac{\mathrm{d}U(z)}{\mathrm{d}z}$$

$$= C_t^{1/2} k(z) \frac{U_S}{z_S}\alpha \left(\frac{z}{z_S}\right)^{(\alpha-1)} \qquad \text{(C. 0. 1-4)}$$

5 Boundary conditions for ground: Under the condition of not considering the roughness, the impact due to roughness shall be corrected by using the exponential relational expression; for the geometrical reproduction of actual building, the boundary conditions adapted to the actual ground conditions shall be used; the logarithmic law shall be adopted for smooth surfaces;

6 Computation rule and spatial description: It is noticeable that periodic unsteady-state fluctuation will occur in the wake flow area of high-rise buildings; the fluctuation is different from turbulence in nature, and it may't be solved by steady-state calculation;

7 Convergence of calculation: The calculate shall be stopped under the condition of full convergence of solution; it shall be determined that the value of the appointed observational point or area does not change, or the root-mean-square residual is less than 10E-4;

8 Selection of turbulence model: The standard k-ε model may be used in case of not high precision and paying attention to the flow field distribution at 1.5m high. When the coefficient of wind pressure on building surface is calculated or high precision is required, the anisotropic turbulence models shall be employed, such as Durbin model or MMK model, etc. ;

9 Array of difference: The array of first-order difference shall be avoided.

Output result:

1) The wind speed at 1.5m height of the pedestrian area around buildings;

2) The magnification coefficient of wind speed in winter: the magnification coefficient of wind speed is required to be less than 2.

C. 0. 2 Building Energy Consumption Simulation

Simulation target:

Firstly, the annual energy consumption of the reference building under the specified operating condition shall be calculated, and then the annual energy consumption of the designed building under the condition of using heat pump type of renewable energy and other energy saving system shall be calculated. When the annual energy consumption of the designed building is not higher than that of the reference building, it meets the requirements. The annual energy consumption of building shall be completed with the help of hourly simulation software. Except the energy storage system consumption calculation and the heat pump type of renewable energy system contribution rate calculation, the other type HVAC system's energy saving calculation may make reference to this standard.

The annual energy consumption of the designed building and reference building shall be conducted according to the following specifications.

Input conditions:

Table C. 0. 2 Setup Parameters of Reference Building and Designed Building

Setup Contents		Reference Building	Designed Building
Thermal Parameters of Envelope		Values shall be taken according to the actual design scheme	Values shall be taken according to *specifications of* DB 11/891 *Design Standard for Energy Efficiency of Residential Buildings* or DB 11/687 *Design Standard for Energy efficiency of Public Buildings of Beijing*
Setup of Operating Conditions	Setup Parameters of Temperature and Humidity for Heating, Venting and Air-conditioning.	Values shall be taken according to the specifications of DB 11/891 *Design Standard for Energy Efficiency of Residential Buildings or* DB 11/687 *Design Standard for Energy efficiency of Public Buildings of* Beijing	
	Fresh Air Quantity	Values shall be taken according to the specifications of DB 11/891 *Design Standard for Energy Efficiency of Residential Buildings* or DB 11/687 *Design Standard for Energy efficiency of Public Buildings* of Beijing	
	Internal Heat Value (Lighting/Indoor People/Equipment)	Values shall be taken according to the actual design scheme. If specific design scheme is unavailable, the values shall be taken according to the specifications of DB 11/687 *Design Standard for Energy efficiency of Public Buildings of* Beijing	
	Outdoor Meteorological Parameters for Calculation	Values shall be taken according to the typical annual meteorological data	
Setup of Heating, Venting and Air-conditioning	Cooling Source System (For different design scheme, the reference systems adopted as shown in the right)	Values shall be taken according to the actual design scheme (In the condition of design adopt water cooled chiller, water-source or ground-source heat pumps, energy storage system)	For the centrifuge and screw machine in electric refrigeration system, its EER and SCOP value shall be taken according to the specifications of DB11/687 *Design Standard for Energy efficiency of Public Buildings* of Beijing
		Values shall be taken according to the actual design scheme (In the condition of design adopt air cooled chiller or evaporation cooled chiller)	For the screw machine in air cooled or evaporation cooled refrigeration system, its EER and SCOP value shall be taken according to the specifications of DB 11/687 *Design Standard for Energy efficiency of Public Buildings* of Beijing
		Values shall be taken according to the actual design scheme (In the condition of design adopt direct expansion system)	Values shall be taken according to the actual design scheme, the efficiency shall meet the requirement about packaged air conditioning unit, multiple air-conditioning (heat pump) unit and ducted air-conditioning unit in DB 11/687 *Design Standard for Energy efficiency of Public Buildings* of Beijing and DB 11/891 *Design Standard for Energy Efficiency of Residential Buildings*

continue C.0.2

Setup Contents		Reference Building	Designed Building
Setup of Heating, Venting and Air-conditioning	Heating Source System	Values shall be taken according to the actual design scheme	Heating source shall adopt gas boiler, the efficiency of boiler shall meet the requirement of DB 11/891 *Design Standard for Energy Efficiency of Residential Buildings*
	Transmission and Distribution System	Values shall be taken according to the actual design scheme	The energy efficiency ratio of transmission and distribution system shall meet the requirement of No. 10.2.1 in this standard
	Terminal Device	Values shall be taken according to the actual design scheme	Values shall be taken according to the actual design scheme

Notices for simulation:

1) The energy consumption for air-conditioning and heating of the reference building and designed building must be calculated by using the same software;

2) The energy consumption for air-conditioning and heating of the reference building and designed building shall be calculated by the typical annual meteorological data.

Output result:

Annual building energy consumption.

C.0.3 Natural Lighting Simulation

Simulation target:

In GB/T 50033 *Standard for Daylighting Design of Buildings*, the standard values of daylight factors for different types of buildings are given. And natural lighting design shall meet both the minimum indoor daylight factor (C_{min}, %) and the critical illumination of interior daylight (lx);

Minimum indoor daylight factor (C_{min}, %): It specifies the minimum daylight factors for different building types and room types.

Critical illumination (lx) of interior daylight: It refers to the illumination of indoor natural light corresponding to the critical illumination of exterior daylight. Different light climatic zones specify the different critical illumination of exterior daylight. Beijing belongs to Class III light climatic zone.

Input conditions:

Since Beijing belongs to Class III light climatic zone, 5000 lx shall be taken as the value of the critical illumination of exterior daylight.

1) Beijing's longitude of 116.317° and latitude of 39.95°;

2) General arrangement drawing of building, specific contour line of building, position of window opening, window form, glass type (refer to the *Standard for Lighting Design of Buildings* for the transmissivity of glass and the reflectivity of indoor floor, ceiling and walls, and it is recommended to consider the surrounding shielding buildings) and indoor house type image; the ceiling height shall be considered for public buildings, and it is recommended to consider the high-rise buildings within the horizontal under angle of 15° as the surrounding shielding buildings;

3) Sky model: CIE Overcast Sky model;

4) Value of critical illumination of exterior daylight: 5000lx;

5) Reference plane: The horizontal plane at 800mm above the indoor floor;

6) Grid spacing: No larger than 1000mm (at least 10 grids are recommended in each direction).

Output Results:

Minimum daylight factor of indoor reference plane;

The distribution of indoor daylighting may be clearly shown on the contour map for the daylight factor of indoor reference plane and the contour map for the critical illumination (lx) of interior daylight of the indoor reference plane.

C.0.4 Natural Ventilation Simulation

There are two methods to simulate natural ventilation according to the different emphasis: one is multizone network simulation method which emphasizes on the overall ventilation of buildings, and it is a lumped model which may be combined with the simulation software for energy consumption of building; the other is CFD simulation method which may describe the characteristics of natural ventilation in a single area. Since both of the methods are used, they are listed in this section together.

1 Multizone network simulation method

Simulation target:

Under the calculating outdoor meteorological conditions (wind speed and wind direction), the natural ventilation air change rate shall be simulated.

Input conditions:

1) Topological graph for building ventilation; the model shall be established on the basis;

2) Resistance model and parameters of ventilation opening;

3) Boundary pressure conditions for opening (it may be obtained according to the outdoor wind environment);

4) If stack ventilation is calculated, indoor and outdoor temperature conditions, indoor heat value and outdoor temperature conditions are needed;

5) Outdoor pressure condition;

6) Introduction on the simplified model.

Output Result:

the natural ventilation air change rate for each room of building.

2 CFD simulation method

Simulation of boundary conditions:

1) Selection of Outdoor Meteorological Parameters

Aiming at this simulation which is used for the studies on indoor air quality for indoor natural ventilation, the representative wind speed and temperature for outdoor simulation are selected, and they are simulated as the steady state.

a) Pressure values of for door and window

The average pressure values for each door and window shall be offered from the simulation results of outdoor wind environment.

b) Outdoor temperature values

The outdoor calculating temperature shall be taken as the outdoor temperature.

c) Relative humidity

Since the impact of relative humidity on indoor air quality only is manifested by temperature rise, the relative humidity is taken only as the condition for judging thermal comfort rather than the boundary condition for simulation.

2) Determination of boundary conditions

It shall be taken as the steady state as well, and the heat release of people, combined floor, roof, orientation of exterior wall and its thermal performancl shall be considered. The boundary conditions shall be determined as follows:

a) Roof: Roof will receive solar radiation and the heating action of outdoor air temperature at the same time. The outdoor comprehensive temperature shall be employed to introduce the temperature rise induced by solar radiation. The outdoor comprehensive temperature shall be calculated by using the formula below:

$$t_s = t_w + \frac{\rho J}{\alpha_w} \qquad (C.0.4\text{-}1)$$

Where, t_s ——Outdoor comprehensive air temperature, ℃;

t_w ——Outdoor design air temperature, ℃;

ρ ——Absorption factor of the envelope external surface to solar radiation;

J ——Total intensity of solar radiation in the orientation where the envelope towards, W/m²;

α_w ——Coefficient of heat transfer for the external surface of envelope, W/(m²·k); 23W/(m²·k) shall be taken;

b) Wall on which the sunshine radiates directly: It shall be treated by the same method as that for roof.

c) Wall on which the sunshine does not radiate directly: Since there is no direct radiation of sunshine, the radiative heat transfer may be ignored. The wall shall be set as constant temperature, and the value of outdoor simulation temperature shall be taken for the outdoor side, whereas the indoor temperature shall be taken for the indoor side.

d) Ceiling: The heat source in the ceiling shall be ignored.

e) Floor or floor slab: When solar radiation is considered, partial floor will absorb heat to result in temperature rise due to the solar radiation incident through window, the floor temperature shall be treated by averaging the solar radiation by the floor area. The heat obtained from the indoor solar radiation incident through glass window shall be calculated by using the formula below:

$$CLQ = FC_s C_n D_{j,\max} C_{LQ} \qquad (C.0.4\text{-}2)$$

Where, CLQ ——heat obtained from the indoor solar radiation incident through glass window;

F ——Net effective area of glass window, m²; it may be calculated by multiplying the window area with the coefficient of effective area (C_a);

$D_{j,\max}$ ——Maximum factor of heat obtained from solar radiation, W/m²;

C_s ——Shielding coefficient of glass window;

C_n ——Shading coefficient of the shading facilities inside window;

C_{LQ} ——Cooling load coefficient.

f) People: As the special boundary, the people's heat value shall be selected according to the actual design scheme or the specifications of DB 11/891 *Design Standard for Energy Saving of Residential Buildings* and DB 11/687 *Design Standard for Energy Saving of Public Buildings* of Beijing.

g) Other objects except equipment which heats, etc. shall be treated as heat insulating boundary.

Notices for simulation:

1) The simulation shall be analyzed according to the steady state;

2) If the interference of the indoor heat source is far larger than the heat transfer of the wall, the heat of the heat conducting part of the wall may be ignored, but the heat obtained from solar radiation shall not be ignored.

Output results:

1) Number of ventilation times for each room of building;

2) Average flow speed of room;

3) Indoor temperature distribution;

4) Distribution of indoor air age.

C.0.5 Outdoor Noise Simulation

Simulation target:

The acoustical simulation shall mainly refer to the requirements of GB 50118 *Code for Design of Sound Insulation of Civil Buildings* and GB 3096 *Environmental Quality Standards for Noise*:

"Noise limitation for functional area of acoustical environment: According to the functional characteristics of regional use and the requirements for environmental quality, the functional zone of acoustical environment may be classified into five levels: Class 0, Class 1, Class 2, Class 3 and Class 4. The Environmental Quality Standard for Noise clearly specifies the limits of environmental noise for the five levels of functional zones, and the limits of noise have become legal standards. It serves as the evidence of judgment in the civil dispute on excessive noise. (it is an article of mandatory rule of law)"

This design specification takes the noise limit for the functional zones of acoustical environment as the standard, the noise map for the functional zones of acoustical environment shall be outputted.

Input conditions: In order to ensure the accuracy of the computer simulation for acoustical environment, the factors of noise source, terrain of simulation area, buildings within the simulation area, etc. shall be inputted, and the specific input conditions are as follows:

1) Building model necessary for simulation analysis in the area coverage;

2) Terrain in the area coverage;

3) Streets, roads, sound barriers, etc. in the area coverage;

4) Report on the monitoring data tested on site for the functional zone of acoustical environment. Since the environmental noise has different impacts on roads of different levels owing to the different traffic flows and different vehicle types passing by, it is recommended to adopt the actually measured more accurate data of traffic noise, or the refer to the data in the relevant standards of *Limits of Noise Emitted by Stationary Road Vehicles*, *Allowable Noise Limits for Motor Vehicle*, *Evaluation About Noise of*

Railway Passenger Cars, *The Limit of Noise Emitted by Rail Bound Locomotives*, *Environmental Quality Standard for Noise*, etc. in the simulation;

5) Report on the monitoring data for noise sensitive buildings in plots.

Output results: Noise in functional zone of acoustical environment

1) The simulation analysis chart of horizontal noise plane (at the height of 1.2m) may clearly show the noise distribution in community;

2) The simulation analysis chart of vertical noise plane (at1m outside the window of building) may clearly show the influence of noise on each position of the building façade.

C. 0. 6 Outdoor Heat Island Simulation

Simulation target:

The distribution of outdoor thermal environment of build may be learnt by the simulation of outdoor heat island of building. It serves as the basis for judging the comfort level of the outdoor microenvironment of building and is used for further instructing architectural design and landscape layout, optimizing planning, construction and landscape schemes, improving the outdoor comfort level, and also reducing the energy consumption and carbon emission of building. In practical works, the reliable program of computer simulation shall be employed to reasonably determine the boundary conditions, and the outdoor thermal environment of building shall be simulated based on typical meteorological condition, so as to reduce the, level of outdoor heat island.

Input conditions:

In order to ensure the accuracy of simulation results, the specific requirements are as follows:

1) Meteorological conditions: The meteorological conditions for simulation may be selected referring to the *Meteorological Data Set Special for Analysis of Building Thermal Environment of China*. It is notable that the meteorological conditions shall cover the parameters of solar radiation intensity, cloud cover in the sky, etc. to be used for the analog computation of solar radiation;

2) Simulation of wind environment: The simulation of outdoor heat island of building shall be established on the basis of the simulation of outdoor wind environment of building to solve various outdoor thermal processes of buildings, so that the intensity of outdoor heat island of building may be realized. Accordingly, the simulation results of outdoor Wind environmental simulation have a direct impact on the calculating results of the intensity of heat island. The simulation of outdoor heat island of building shall meet the requirements for the simulation of outdoor wind environment of building. It includes computational domain, model reproduction domain, requirements for gridding, boundary conditions for inlet, boundary conditions for ground, computation rule and convergence, array of difference, turbulence model, etc. ;

3) Simulation of solar radiation: In the simulation of outdoor heat island of building, the simulation of solar radiation for building surface and underlying surface is an important link of simulation and an important influencing factor for the intensity of outdoor heat island. The simulation of solar radiation shall take direct solar radiation, scattered solar radiation, multiple reflective radiations and long-wave radiations among surfaces, etc. into consideration. In practical application, appropriate simulation software shall be used; if the factor of radiation computation or scattered computation for the part of

multiple reflectivity is not considered in the employed software, the simulation results shall be corrected to satisfy the requirement for accuracy of analog computation;

4) Parameter setting for underlying surface and building surfaces: For building surfaces and underlying surface, the material parameters of physical properties, reflectivity, permeability, evaporation rate, etc. shall be set so as to accurately compute the solar radiation and heat transfer on building surfaces and underlying surface;

5) Parameter setting of landscape factors: In the outdoor thermal environment of building, the landscape factors of plants, water body, etc. have great impact on the simulation result, thus relevant setting is required in the simulation. For plants, the influence of plants on wind environment may be simulated according to the theory of porous medium, and the influence of plant on thermal environment shall be calculated according to the computation of heat balance for plants and based on the results of radiation computation and the rate of evaporation rate, etc., so that the influence of plants on outdoor microenvironment of building may be completely reflected. For water body, different settings may be performed for static water surface and fountain. The above settings may be appropriately simplified in engineering application.

Output results:

The outdoor heat island intensity may be simulated to obtain the outdoor temperature distribution of building, and the computation results of the average intensity of outdoor heat island may be given to support the landscape design of building. However, in order to verify the accuracy of simulation, the simulation results of accumulated solar radiation for all surfaces, computation results of surface temperature for building surfaces and underlying surface, the simulation results of outdoor wind environment of building, etc. shall be provided at the same time.

Explanation of Wording in This Standard

1 Words used for different degrees of strictness are explained as follows in order to mark the differences in executing the requirements in this standard:

    1) Words denoting a very strict or mandatory requirement:

      "Must" is used for affirmation; "must not" for negation.

    2) Words denoting a strict requirement under normal conditions:

      "Shall" is used for affirmation; "shall not" for negation.

    3) Words denoting a permission of a slight choice or an indication of the most suitable choice when conditions permit:

"Should" is used for affirmation; "should not" for negation.

    4) "May" is used to express the option available, sometimes with the conditional permission.

2 "Shall comply with..." or "Shall meet the requirements of..." is used in this code to indicate that it is necessary to comply with the requirements stipulated in other relative standards and codes.

List of Quoted Standards

GB 3096　　Environmental Quality Standards for Noise

GB 6566　　Limits of Radionuclides in Building Materials

GB 18580　　Indoor Decorating and Refurbishing Materials - Limit of Formaldehyde Emission of Wood-based Panels and Finishing Products

GB 18588　　Limit of Ammonia Emitted from the Concrete Admixtures

GB 18613　　Minimum Allowable Values of Energy Efficiency and The Energy Efficiency Grades for Small and Medium Three-phase Asynchronous Motors

GB/T 18870　　Technical Conditions for Water Saving Products and General Regulation for Management

GB/T 19939　　Technical Requirements for Grid Connection of PV System

GB 20052　　Energy Efficiency and the Evaluating Values of Energy Conservation for Three-phase Distribution Transformers

GB 50015　　Code for Design of Building Water Supply and Drainage

GB 50033　　Standard for Daylighting Design of Buildings

GB 50034　　Standard for Lighting Design of Buildings

GB 50057　　Design Code for Protection of Structures Against Lightning

GB 50068　　Unified Standard for Reliability Design of Building Structures

GB 50118　　Code for Design of Sound Insulation of Civil Buildings

GB 50180　　Code of Urban Residential Areas Planning & Design

GB 50217　　Code for Design of Cables of Electric Engineering

GB/T 50314　　Standard for Design of Intelligent Building

GB 50325　　Code for Indoor Environmental Pollution Control of Civil Building Engineering

GB 50352　　Code for Design of Civil Buildings

GB/T 50378　　Evaluation Standard for Green Building

GB 50400　　Engineering Technical Code for Rain Utilization in Building and Sub-district

GB 50555　　Standard for Water Saving Design in Civil Building

GB 50736　　Design Code for Heating Ventilation and Air Conditioning of Civil Buildings

GB 50763　　Codes for accessibility design

SJ/T 11127　　Overvoltage Protection for Photovoltaic (PV) Power Generating Systems - Guide

CJ 164　　Domestic Water Saving Devices

CJT 174　　Technological Classification of Community Intellectualization System

TSG G0002　　Supervision Administration Regulation on Energy Conservation Technology for Boiler

JC/T 945　　Water Permeable Brick

JGJ 16　　Code for Electrical Design of Civil Buildings

JGJ/T 154　　Standard for Energy Consumption Survey of Civil Buildings

JGJ/T 163　*Code for Lighting Design of Urban Nightscape*

JGJ 203　*Technical Code for Application of Solar Photovoltaic System of Civil Buildings*

DB 11/T 211　*Plant Materials for Urban Landscape Greening Tree Seedling*

DB 11/T 281　*Code for Roof Greening*

DB 11/ 343　*Technical Specification for Water Saving Apparatus*

DB 11/T 388.1-4　*Technical Code of Urban Nightscape Lighting*

DB 11/T 624　*Technical Requirements for Measurement of Electricity Usage by Category in Office Buildings of Public Agencies*

DB 11/T 686　*Specification for Construction and Acceptance of Water Permeable Brick Pavement*

DB 11/687　*Design Standard for Energy Efficiency of Public Buildings*

DB 11/T 825　*Evaluation Standard for Green Building*

DB 11/891　*Design Standard for Energy Efficiency of Residential Buildings*